Geography as Spatial Interaction

Geography as Spatial Interaction

Edward L. Ullman

Edited by
Ronald R. Boyce

Foreword by
Chauncy D. Harris

UNIVERSITY OF WASHINGTON PRESS
Seattle and London

Library of Congress Cataloging in Publication Data
Ullman Edward Louis, 1912–1976.
  Geography as spatial interaction.

  1. Geography. I. Boyce, Ronald R. II. Title.
G116.U44   910   79-6759
ISBN 0-295-95711-5

Designed by Veronica Seyd
The decorations on the case and the part-title pages are some char-
acteristic doodles selected from numerous examples found among
Edward L. Ullman's papers.

Foreword

A number of years ago my long-time good friend Edward L. Ullman shared with me his hope of bringing his papers together into a collected volume. I was delighted with the idea since it meant that his seminal writings, scattered through many sources and published in diverse places—articles in national periodicals, mimeographed reports (called fugitive materials by librarians), chapters in books or government documents, and papers that were still in manuscript form—could become more widely available. In 1970 he invited me to write a foreword to the volume. I was honored and pleased as at various times in our respective careers we had been very close personally and intellectually. Neither of us at that time had any idea it might be a posthumous publication.

The written word does not catch the whole flavor of his personality, yet it is important that his most significant nonmonographic writings be gathered in one publication, in which his principal ideas, concepts, suggestions, and findings can be conveniently read, digested, consulted, and enjoyed. Ronald R. Boyce, of Seattle Pacific University, is to be thanked for undertaking the task of editing this volume. He was a student, collaborator, and colleague of Edward Ullman at the University of Washington.

In 1977 John D. Eyre, of the University of North Carolina at Chapel Hill, edited a volume of eight essays that appraise Ullman's contributions to various fields of geography and discuss the present status of these fields. In *A Man for All Regions: The Contributions of Edward L. Ullman to Geography*, Richard L. Morrill, of the University of Washington, treats geography as spatial interaction; Clyde E. Browning, of the University of North Carolina at Chapel Hill, discusses core and fringe in regional development; Wilbur Zelinsky, of Pennsylvania State University, revisits the amenity factor; Norton Ginsburg, of the University of Chicago, examines geography and economic development; Chauncy D. Harris summarizes patterns of cities; Arthur Getis, of the University of Illinois at Urbana, evaluates the urban economic base; Ronald R. Boyce treats applied geography in regional development; and Harold M. Mayer, of the University of Wisconsin at Milwaukee, examines the changing American railroad pattern.

The critical informed evaluations by leading scholars in each field in *A Man for All Regions* provides a supplement to this collection of Ullman's actual writings. John Eyre himself gives a splendid account of Ullman's life, "Edward Ullman: A Career Profile." In that same volume, Joan Connelly Ullman, of the History Department, University of Washington, who shared the last nine years with Edward Ullman, has provided biographical data including a detailed list of his professional positions and activities, publications, talks, conferences attended, and principal travels.

A number of other memorials of Ullman have appeared in leading international journals: *Annales de géographie* (Raymonde Caralp, 1976) *Bollettino della società geografica italiana* (Carlo Della Valle, 1976); *Annals of the Association of American Geographers* (Chauncy D. Harris, 1977).

One measure of a person's impact on a field is the number of citations of his publications. The cumulative indexes to the *Geographical Review* record fifty-seven references to Ullman, extending over the years 1937–75. The cumulative index for the *Annals of the Association of American Geographers* for the years 1936 to 1965 records twenty-two references to Ullman.

In view of the current interest in biographies of major contributors to geography, it may be appropriate to suggest individuals whose contact with Ullman was particularly exciting to him at various periods in his career. The professors he found most congenial and stimulating were Charles C. Colby in urban geography, with whom he wrote his doctoral dissertation; Robert S. Platt in transportation geography and the concept of regions of organization; and Louis Wirth in urban sociology, all at the University of Chicago. Derwent S. Whittlesey of Harvard University turned Ullman toward political geography, particularly that of boundaries. In this early period of interest in boundaries he and Stephen B. Jones were particularly congenial. As fellow graduate students, Ullman and I developed an early—and sustained a very active—intellectual dialogue. The single author he read with greatest excitement was doubtless Walter Christaller.

During World War II in Washington, D.C., he worked closely with Richard Hartshorne, Preston E. James, and Edward A. Ackerman. During his early career at Harvard he engaged in lively discussions with Walter Isard on interregional relations, out of which they eventually presented a joint paper. At the University of Washington he acted closely with many colleagues and students, particularly Michael F. Dacey, Ronald R. Boyce, and John D. Eyre. He also carried on an extensive correspondence with a wide range of geographers, scholars in other disciplines, and men of affairs, as happily recorded in excerpts in this volume. That he was in close touch with, and respected by, leaders in the discipline of geography in his generation is suggested by the fact that eight of these individuals with whom he had the most intimate intellectual interaction served as president or honorary president of the Association of American Geographers.

Certain qualities of Edward L. Ullman stand out particularly in my memory of our association over forty-three years. First and foremost was the scintillating quality of his conversation, with its stream of interesting ideas and probing suggestions, and his delight in trying out new ideas. Dialogue with him was lively and fruitful. His second quality was his search for generalizing principles or regularities. How did a particular phenomenon or observation fit into some larger pattern? Thirdly, he was rich in comparisons of concepts from diverse fields. And finally, his discussions were larded with apt examples and concrete illustrations drawn from his keen observations in the many places he had visited, or from his readings and conversations.

MAJOR CONTRIBUTIONS

In my judgment Edward Ullman's most significant contributions revolved around three key elements: the city, transportation, and regional development. Often his work involved two or three of these elements simultaneously, as did his lifelong interest in the relations of cities to their supporting areas, especially through actual movements of goods and services, first developed in his doctoral dissertation on Mobile, Alabama.

In urban geography his major contributions included analyzing general patterns in the distribution of cities, especially of central places and transport cities (he was the first American geographer to address this problem systematically); and examining the internal patterns of cities and the basic and nonbasic elements in the economic structure of American cities, especially through the approach of minimum requirements.

His interest in transportation was expressed in the definition of geography as spatial interaction. He analyzed the bases of spatial interaction in complementarity, transferability, and intervening opportunity. He studied American railroads and produced the first summary map of railroad capacity and later of traffic flows on American and Canadian railroads. In a substantial research project he analyzed the interstate movements of commodity groups in the United States by rail, by Great Lakes ships, and by coastal shipping.

In extended attention to problems of regional development, Ullman emphasized amenities as a factor in growth. He considered spatial aspects of underdevelopment at various scales, and noted the complexity of factors in development and the urban-rural differences in development in Southeast Asia. As director of the Meramec Basin Research Project he proposed, as crucial for the regional development of that area, the construction of a reservoir for recreation.

He touched a variety of other fields, from his early interest in the political geography of boundaries to the contributions of geographers to area studies, trade-offs between time and space, air pollution in metropolitan areas, and finally the problems of Seattle and the Pacific Northwest and the role of the University of Washington in them.

I will mention some of his important publications with emphasis on the more seminal papers and on the monographs or papers that are not reproduced in this volume. These summaries, however, are only a hint of the richness of the original publications.

DOCTORAL DISSERTATION

On rereading Ullman's first monograph, his doctoral dissertation, "Mobile: Industrial Seaport and Trade Center" (1943), having first read it in manuscript some thirty-six years ago, I was struck initially by the superb "Character of the City." This brief introduction was a model of descriptive characterization later reproduced in a syllabus in English at the University of Chicago as an illustration of expository writing. I was also impressed once again by the balanced treatment of evolution, site, internal patterns, supporting areas, and transportation of the city. The study was particularly distinguished for its innovative analysis of the supporting area (hinterland), the effect of transportation on Mobile, and the nature of the currents of traffic that support the city, reflecting Ullman's special interest in transportation. He measured the flows of traffic by individual railroad lines, highways, water routes, natural gas pipelines, and high-voltage electricity transmission lines, inbound and outbound; and analyzed the structure of these flows and their sources and destinations in one of the first attempts at a comprehensive understanding of the role of transportation and traffic flows in the life of a city and its ties to its region. Mobile is illuminated by comparisons with other cities.

URBAN GEOGRAPHY

Reflecting on the spatial patterns of the cities of Iowa and on the need for some theory or typology to describe and explain them, Ullman was evolving a sort of central place theory, when in 1938 at Harvard he talked with August Lösch, who told him of Walter Christaller's work on central places in Southern Germany. In 1938–39 when we were graduate students at the University of Chicago, we discussed this theory frequently. He later wrote his pioneer article "A Theory of Location for Cities" (1941), reproduced in chapter 11 of this volume. This paper was the first full, clear statement of central place theory in English, succinctly summarizing the contribution of Walter Christaller, but adding significantly the contribution of others and including many of Ullman's original ideas.

"The Nature of Cities" (1945), which we wrote together on weekends while in military service during World War II, addresses questions of the distribution of cities and of the internal patterns of cities. In "The Nature of Cities Reconsidered," his presidential address to the Regional Science Association in 1962, Ullman analyzed the internal expansion of urban areas, with emphasis on the role of improved circulation and communication, particularly as a result of the development of the automobile and improved highways (see chap. 12).

In "The Minimum Requirements Approach to the Urban Economic Base" (1960), Ullman and Michael F. Dacey used occupational data from a fourteen-industry classification (U.S. Census of Population) to separate from the total labor force that portion in each sector of the economy needed to maintain the city itself, that is, those people engaged in local service or nonbasic activities. The percentage engaged in export activities from the city, or basic employment, could then be estimated.

In each of the census categories the proportion of the labor force serving local needs was approximated by the lowest figure for any city in a particular size group in the country, for example, the percentage occupied in durable manufacturing in Washington, D.C., among cities of more than a million population. The minima for each of the fourteen industries were then summed, and this figure was considered a useful measure of the total employment in local service. The calculations revealed that the proportion of the total labor force in a city serving local needs increased with the size of city, from 24 percent in cities of 2,500 to 3,000 population up to about 57 percent for metropolitan areas of more than a million population. It was noted that large cities can provide many specialized services internally which must be sought from outside by small centers.

Ullman and Dacey suggested that minimum requirements could be used to estimate the ratios of basic to nonbasic activities in each of the fourteen industries, thus producing a preliminary profile of basic and nonbasic activities in any given city. The method was also found to have utility in analyzing city types. The use of minimum requirements rather than a national mean or median had been suggested earlier, but Ullman and Dacey developed the method more rigorously and suggested most effectively its significance.

This approach had earlier been developed by Ullman in "Sources of Support for the San Francisco Bay Area Economic Base" (1959). Challenged by a critical note on this method, Ullman several years later restated its advantages in an article entitled "Minimum Requirements After a Decade: A Critique and an Appraisal" (1968).

In the monograph *The Economic Base of American Cities: Profiles for the 101 Metropolitan Areas over 250,000 Population Based on Minimum Requirements for 1960* (1969), Ullman joined with Michael Dacey and Harold Brodsky in a study applying comparative methods to an analysis of the basic support of the 101 largest cities of the United States, utilizing 1960 data. The detailed data are included as an appendix, and are not textually analyzed except in discussions of comparisons with other methods or of regression coefficients for differences in the size of cities. Comparisons with other general methods, such as the location quotient, were probed more deeply than previously. The authors also compared detailed surveys of individual cities: Washington, D.C.; Peoria, Illinois; San Francisco; and Los Angeles. They concluded that generally the minimum requirements approach represented an inexpensive and satisfactory estimate of the basic employment support of a city and a rough measure of the multiplier effect of additional basic employment.

TRANSPORTATION GEOGRAPHY

In "The Railroad Pattern of the United States" (1949), Ullman classified railroads on the basis of number of tracks or of signaling equipment (which increases capacity). This was a pioneering study in its attempt to utilize quantitative measures of the pattern of facilities and freight traffic.

Ullman coined the phrase "geography as spatial interaction" during a meeting of the Western Political Science Association in Pullman, Washington, in 1952. A speaker preceding him had defined sociology as the study of social interaction; Ullman reflected that geography could then be defined as the study of spatial interaction, an idea that he developed in two similar papers, "Geography as Spatial Interaction" (1954) and "The Role of Transportation and the Bases for Interaction" (1956). A revised version of these papers appears in this volume (chap. 1).

With Harold M. Mayer, Ullman wrote the chapter "Transportation Geography" in *American Geography: Inventory and Prospect* (1954). In 1956 he published an innovative map of traffic flow on the railroads of the United States, showing traffic in millions of tons on class one railway lines. The map first appeared in "The Role of Transportation and the Bases for Interaction" (1956), but its analysis awaited the publication of *American Commodity Flow* (1957).

The monograph *American Commodity Flow: A Geographical Interpretation of Rail and Water Traffic Based on Principles of Spatial Interchange* represents one of Ullman's most sustained research efforts. He compiled summary maps of freight movements in the country by rail, Great Lakes ships, and coastal shipping. He noted particularly that the principal freight flows are east-west, though the physical grain of the country is north-south. He also delineated major freight areas: the industrial belt of the Northeast as the destination of most raw material and fuel shipments and the source of most manufactured products; and other special regions of freight generation, such as the corn and wheat belts west of the industrial belt, the forest regions of the Pacific Northwest and the South, and the specialized off-season fruit and vegetable areas of California and Florida. He discussed the commodities with very heavy freight flows: coal from the Central Appalachians by rail, by Great Lakes ships, and by coastal movement on the Eastern Seaboard; iron ore through the Great Lakes; oil by pipeline and tanker, and natural gas by pipeline, these latter two over long distances from southwestern fields into the industrial belt. In an analysis of railroad offices and yards he noted the gateway location of headquarters for many railroads (whether of ports on the Eastern Seaboard or on the southern and western margins of the industrial belt); the location of traffic solicitation offices where freight is generated, generally corresponding to size of city, but with Detroit (autos) and Pittsburgh (steel) standing out; and the distribution of the largest rail classification yards athwart the main flows of freight at entrances to or exits from the steeper grades over the Appalachians. The bulk of the monograph, however, consisted of the analysis of freight flows from or to twenty representative states by some or all of six commodity groups, based on the Interstate Commerce Commission's 1 percent sample, *Carload Waybill Statistics*, with more detailed illustrative discussions of Iowa as a farm state, Connecticut as an industrial state, and Washington as a producer of forest products. (This section was also published in German in *Die Erde* [1955].) The final chapter summarized American ocean freight traffic, both domestic and foreign, in a brief but effective synthesis. The study as a whole is rich in detailed maps but spare in text.

Ullman was concerned with rivers as linkages. One of his most interesting and original papers was "Rivers as Regional Bonds: The Columbia-Snake Example" (1951), in which he proposed the term "dioric" for stretches of streams flowing across mountains. He then analyzed the Columbia-Snake river system in terms of two dioric sections through mountains and two exotic sections across desert stretches, and thereby demonstrated the utility of the concepts. This article is reprinted in chapter 7.

In 1956 Ullman spent a half year as transportation consultant to the Stanford Research Institute and member of a mission which prepared *An Economic Analysis of Philippine Domestic Transportation* (1957). Ullman himself prepared sections on forest and mineral products, manufacturing, and trade centers. Based on this work he also published an article "Trade Centers and Tributary Areas of the Philippines" (1960), an analysis of the hierarchy of central places and the delimitation of their hinterlands based in part on the study of traffic patterns on the roads (republished here in chap. 8).

REGIONAL DEVELOPMENT

The versatility and breadth of Ullman's mind and interests are revealed by two contrasting articles: one on amenities as a factor in diffusion of economic activities and population to the West and South of the United States, and the other on factors favoring industrial concentration in the Northeast. "Amenities as a Factor in Regional Growth" (1954) discussed the rapid growth in population of California and Florida and the attractive-site factors in large-scale migration to them (chap. 9). On the other hand, "Regional Development and the Geography of Concentration" (1958) emphasized the concentration of development in a core area of the manufacturing belt of the United States, based on assembly and market advantages of this area (chap. 5). This paper was also published in Italian (1959).

"Geographic Theory and Underdeveloped Areas" (1960) examined the spatial aspects of underdevelopment, noting that within many countries there are sharp contrasts between well-developed and less-developed regions and, in a yet finer areal breakdown, contrasts among adjacent villages (chap. 6).

In "The Primate City and Urbanization in Southeast Asia: A Preliminary Speculation" (1968), Ullman noted the preponderance of primate cities in the countries of Southeast Asia, and speculated about their advantages—primarily economic efficiency—and disadvantages—potential political alienation of an urbanized elite from the rural segment of the country.

For two years, 1959–61, Edward Ullman was director of the Meramec Basin Research Project, which made a careful study of possible water and economic development of an area extending about one hundred miles southwest from St. Louis into the Ozark Highlands. He prepared a three-volume report, *The Meramec Basin: Water and Economic Development* (1962), in collaboration with Ronald R. Boyce and Donald J. Volk. The study was distinguished methodologically as evident in "Geographical Prediction and Theory: The Measurement of Recreation Benefits in the Meramec Basin" (1967), which is included in revised form in chapter 10.

OTHER CONTRIBUTIONS

Ullman's two earliest published papers, "The Historical Geography of the Eastern Boundary of Rhode Island" (1936) and "The Eastern Rhode Island-Massachusetts Boundary Zone" (1939), reflected two years of graduate work at Harvard University. Both articles revealed early scholarly capabilities and productivity, but he published no more papers in political geography.

In "Human Geography and Area Research" (1953), he discussed some of the special contributions that geographers could make to interdisciplinary area studies (chap. 2).

In "Problems of the Industrial Landscape: The City and Environmental Quality, Especially Air Pollution Sources and Costs," delivered at a 1971 UNESCO Conference on Man and the Biosphere and published in 1974, he noted that air pollution should be approached as a problem of the *concentration* of industry, transportation, and population in large cities (chap. 13).

During his last years Ullman was much concerned with the trade-offs between time and space, as recorded in his last major publication, "Space and/or Time: Opportunity for Substitution and Prediction" (1974), a paper delivered at an annual conference of the Institute of British Geographers in 1973 (chap. 3).

The contributions of Edward Ullman have been highly esteemed and widely recognized by leaders in the discipline of geography and in related fields. In 1958 the Association of American Geographers awarded him a Citation for Meritorious Contribution to the field of geography "for his quantitatively conceived and imaginative investigations in the geography of transportation and spatial interaction," and in 1972 named him Distinguished Geographer. The Italian Geographical Society honored him as a Corresponding Member and in 1959 awarded him a Citation for Meritorious Contributions. The Regional Science Association elected him president in 1961.

Edward L. Ullman was above all an idea man, stimulating in conversation and stimulated by lively dialogue, interested primarily in concepts illustrated by concrete examples, open to and conversant with concepts from related social sciences, and engaged from time to time intensively in applied geography. His ideas found expression primarily in scattered articles, papers, and talks rather than in monographs. This collection, *Geography as Spatial Interaction*, renders scholars a great service by gathering into convenient, accessible form many of Ullman's essential writings.

CHAUNCY D. HARRIS

Preface

For over a decade, Edward L. Ullman had been discussing the possibility of publishing in book form a collection of his articles, but at the time of his death, 24 April 1976, this project was still far from completion. He had selected the title, "Geography as Spatial Interaction," and had written a brief introduction. He had also prepared a tentative table of contents, listing the articles to be included organized under appropriate subtopics. He had intended to update the articles by adding footnotes with appropriate commentary, but this plan was never realized.

In preparing this collection for publication, I have attempted to remain true to Ullman's original intention. The title and the introduction are his. I have, however, added to his original introduction several brief statements that illuminate his approach to geography and his criteria for the evaluation of his own works. As far as possible I have retained his organizational structure. Primarily for reasons of space, however, coauthored articles have been omitted, as have a few pieces that seemed dated or were of primarily local concern.

I tried at first to add commentary in footnotes as Ullman himself had planned to do, but this proved to be impossible, since I found that I was inserting my own opinions and comments rather than his. I next tried to use selected comments from his voluminous correspondence as footnotes at various places in the articles, but this too presented a problem, since it was not always possible to find an appropriate comment that would speak to the specific point being raised. The solution I arrived at was to select excerpts from Ullman's correspondence that related to each of the articles and use them to introduce the article in question. These excerpts appear in the margins at the beginning of the article, following my own brief commentary, which provides the circumstances of its composition and publication. While the excerpts presented in this way can certainly not be considered a satisfactory substitute for the comments Ullman would have written for this purpose, they do provide some insight into his thoughts about his own writings, before, during, or after the time they were actually produced.

The articles are reprinted here as they were published, except for the correction of typographical errors or errors noted by Ullman on his copies, and minor modifications made in subheads, footnotes, and the like for typographical consistency. For previously unpublished manuscripts, the latest version has been used.

The decision not to include coauthored papers has resulted in the omission of several articles that presented concepts almost synonymous with Ullman's name, such as the multiple nuclei theory and the minimum requirements method. The one partial exception to this rule is chapter 10, "Geographical Prediction, Regional Planning, and the Measure of Recreation Benefits in the Meramec Basin," for which the technical measurement aspects were worked out primarily by

Donald Volk. It is the spatial analog principle, however, that best reveals Ullman's approach to the problems posed in this paper. As he says in his Introduction, quoting Herbert Lundy, he is interested not so much in method or technique as in "catching fish."

Ullman was a powerful writer who wrote in simple and exciting prose. It is an indication of the breadth of his interests that many of these papers were published, not in the *Annals of the Association of American Geographers* but in journals associated with other disciplines, both in this country and abroad. Nevertheless, the title of this book expresses a unifying theme: the significance of spatial interaction in the understanding of man and his activities over the surface of the earth. In the lead article, "Geography as Spatial Interaction" (chap. 1), he discussed in detail the force of situation, referring to "the effects of one area, or rather phenomena in one area, on another area." This force, though almost impossible to measure, established all manner of spatial interaction; and in his work on American railroad patterns he observed its results, noting that it caused routes to run "cross-grain" to the natural physical channels of the nation. Thus he demonstrated that the situational force was far more powerful than the physical, or site, features of the earth.

Almost all of Ullman's work involves the identification and measurement of interaction among complementary regions. His discussion of central place theory, for example, centers about city-rural interaction. The port-outport study clearly deals with the matter of spatial interrelationships. Even his last major work on time and space, in its treatment of periodic centers, interprets the interplay between areas. Thus it seems fitting that this book, which brings together a lifetime of geographic investigation, should bear the title *Geography as Spatial Interaction*.

In the Introduction I have included four short documents, starting with the brief statement Ullman wrote specifically for this purpose, in which he clearly articulates his approach to geography. "Recent Stages in Geographic Thought," from a talk he used to give to graduate students, summarizes the history of geography from 1900 to the present. In "The Contribution of a Geographer" and "The Test of Significance in Geography," both previously unpublished papers, he sets forth the criteria by which his own works can be evaluated. The marginal extracts drawn from his correspondence include pertinent comments about his ideas for this book.

Part I includes articles that focus on "the nature of geography." It begins with a 1954 version of an unpublished paper, "Geography as Spatial Interaction," which had a tremendous impact on many of us graduate students at the University of Washington in the late 1950s, and concludes with Ullman's final major paper, "Space and/or Time" (chap. 3), which culminates a lifetime of concern about the dominance of situation versus site in the explanation of things spatial. In this paper, time is used as "a motor for situation concepts"—something for which he had been searching for years. Here, by giving operational power to "situation" by means of harnessing transportation to time, Ullman has opened up for us all many new avenues of research.

In Part II, various articles dealing with regions and regionalism are presented. It will be quickly apparent that Ullman's approach to regions was far different from that customarily used in regional geography. His regions are dynamic, interacting with each other, always pieces of a greater whole. They were defined in terms of "situation" (horizontal relationships) rather than "site" (vertical relationships).

The first article, "Regional Structure and Arrangement" (chap. 4), is previously unpublished, although ideas drawn from it appear in *American Geography: Inventory and Prospect* (Syracuse, N.Y.: Syracuse University Press, 1954). The topic was also further explored in the article "Regional Development and the Geography of Concentration" (chap. 5), which was presented as a paper at a Friday symposium when I was a graduate student at the University of Washington and was my first personal introduction to Edward Ullman. The principal theme, developed with fascinating variations, extensions, and applications, is that great regional disparity is found within areas, of such a serious nature that normative data pertaining to them reflect only "meaningless averages." The secondary point made is that only a few places on earth show concentrations of major development; the conclusion is that there is underdevelopment almost everywhere. The application of this theme is found in "Geographic Theory and Underdeveloped Areas" (chap. 6), in which Ullman deals with some of the problems of underdevelopment.

Part III includes articles demonstrating Ullman's applications of his theories to particular places: the Pacific Northwest (chap. 7), the Philippines (chap. 8), and the Meramec Basin in Missouri (chap. 10). Also found here is the classic work on "Amenities as a Factor in Regional Growth" (chap. 9). Ullman was always pleased to label himself an applied geographer, and in many respects these papers verify the significance of his pragmatic approach to geography by the solving of real problems. He was, in addition, a geographer's geographer, who pioneered for the profession the practical value of geography among the social sciences.

Part IV deals with cities, one of Ullman's favorite topics, along with spatial interaction and the nature of situation. The papers here presented reflect his multifaceted treatment of cities: first, as simply points on the surface of the earth, and then as areal units with their unique qualities of interaction. The publication of the single article "A Theory of Location for Cities" (chap. 11) would be sufficient to assure him an important place in the historical development of American geography. This seminal work on central place theory, although modestly presented as a summary translation of Walter Christaller's *Die zentralen Orte*, nonetheless contains many original insights and applications, some of which have not been fully followed up to this day. Attached to this article is a formerly unpublished addendum which Ullman and I prepared together.

Chapter 12, "The Nature of Cities Reconsidered," was Ullman's presidential address to the Regional Science Association. The portion of this article that discusses some of the interplay of "site" and "situation" in an intraurban context has a highly personal meaning for me. I recall the day in 1961 when we first discussed this idea in the Faculty Club at Washington University in St. Louis. Ullman was excited and elated and could hardly wait to talk about the applications with others in the room. At the time I thought that I had generated the concept, but I have come to realize that I had simply triggered an idea that he had been thinking about in a slightly different form for many years.

"Problems of the Industrial Landscape: The City and Environmental Quality, Especially Air Pollution Sources and Costs" (chap. 13) reflects Ullman's astute awareness of and concern with the quality of urban life. He had a considerable interest in aesthetics, noise pollution, and many other items of community concern. He frequently wrote letters to the editors of the local newspapers and attempted in other ways to upgrade the urban quality. This article reflects the degree of his commitment to such practical matters and is a good example of the thoroughness with which he attacked almost any problem.

The final article, "City, Port, and Outport" (chap. 14), is unfinished. His service as a board member of Amtrak, which he took very seriously; his many other duties; and his health prevented him from completing this article. The ideas presented are sufficiently clear, however, to make it possible that someone else may some day take up the study where he left off.

I wish to thank the many persons who have made the publication of this posthumous work possible. In particular, I am grateful to Professor Joan Connelly Ullman for allowing me to ferret out and use, without restrictions, comments from Edward Ullman's letters, which are now in the archives of the University of Washington.

Professor Chauncy D. Harris of the Center for International Studies, University of Chicago, has been my confidant and guide throughout the work. He has been of inestimable aid and has provided crucial editorial advice. In addition to writing the Foreword for this volume, as Ullman had requested, Professor Harris compiled the bibliography, "Selected Writings of Edward L. Ullman," which appears herein.

Funds for photocopying materials were graciously provided by Dean Morgan Thomas, director of the Joint Center for Urban Studies at the University of Washington. Most of the excerpts from letters were originally transcribed by the secretarial staff of the University of Washington's Geography Department.

I also wish to thank members of the staff of the School of Social and Behavioral Sciences at Seattle Pacific University. Dorothy Wilson, administrative secretary of the school, has overseen the burdensome task of typing the manuscript and has been constantly involved in the logistical advancement of the work. My daughter, Renée, performed all the initial work of article search and presentation. Final typing and proofing were done by Beth Howard, Meleani Hutton, Janis Tucker, and Winifred Sheard.

Special thanks are due Mrs. Anna Chiong, director of the University of Washington Geography Library, for her aid in finding the many incomplete and unclear references in the original articles.

Finally, I wish to thank the original publishers of these articles for permission to reprint them in this volume. The evolution of each paper, as far as I have been able to determine it, is given in the marginal note for that paper. Acknowledgment is made here to the *American Journal of Sociology*, the *Annals of the Association of American Geographers*, the *Geographical Review*, the Institute of British Geographers, the University of Chicago Press, and Verlag Dokumentation, K. G. Saur Publishers, Munich, and to Basic Books, Inc. for the use of figures 1, 3, 4, 5, and 6 in chapter 10.

Full responsibility for any errors is mine. I also take responsibility for the selection of the comments from Ullman's unpublished letters, for the final choice of papers to be included or omitted, and for the general organizational structure insofar as it differs from that originally proposed by Ullman himself.

RONALD R. BOYCE
October 1979
Seattle, Washington

Contents

Geography
as
Spatial

Interaction

Introduction

Whatever virtue these studies have is related in part to their being generally based on an idea—an idea not a technique, but an idea applied to a real body of data. So much of current research is an attempt to exploit and develop techniques, that they are applied either to hypothetical or to inadequate bodies of data.

A second aspect of the studies in this book is that they are aimed at findings, not theories or methods. As Herbert Lundy says, "I do not place the method before the results—I like to catch fish." In a sense they attempt to build a shack with a roof which at least will enable one to exist, as opposed to a solid foundation which is useless without a roof, which may never be built. As Hettner says, we are not interested in theory for theory's sake, but rather for its contribution.

In the process of building theories which provide roofs, however, I am conscious of the need for pushing techniques further in order to make the theory more generally applicable. In the Meramec study all that was needed was a breakthrough invention to predict attendance at one reservoir, but to make the theory more generally applicable the intervening opportunity problem should be solved. (Here I am quite certain it can be done, but the mathematics are complicated.) What I am saying is that in addition to applying techniques to important ideas, we also need even better techniques. Therefore, I can sympathize with slow progress, and the need for developing foundations. However, more stress needs to be given to the assumptions and the geographical peculiarities of data that go into tests of theories and techniques, rather than to the workings of the technique or model itself.

Recent Stages in Geographic Thought

Before demonstrating the case, let us consider the recent history of geography. Geography has been preeminently a descriptive science and thus has hardly attempted prediction—much though we need to do this, and much though I hope to demonstrate that it is possible.

The recent history of geography in America might be divided into three overlapping periods very roughly as follows: 1900 to 1920, 1920 to 1945, and 1945 to the present.

The first stage, 1900 to 1920 or 1925, emphasized the relations between man and his environment. This was essentially a neo-Darwinian reflection in geography, just as occurred in other fields. It was criticized as being too deterministic—of making too much of the environment, of over-asserting and over-simplifying environmental controls over man's action.

As a reaction to this environmental determinism, a second general approach, a new overlapping stage, developed from roughly 1920 or 1925 to 1940 or 1945. This new emphasis actually went back to an even earlier conception of geography, essentially more descriptive, but emphasizing the *differences* from place to place and was eloquently and exhaustively documented in Richard Hartshorne's justly famous *The Nature of Geography* (1937).[1] A shorthand way of describing or defining geography was to define it as the science of areal differentiation. This stage coincided particularly with regional approaches and regional descriptions. Some critics remarked that if the previous period had been determinism, this second period might be even worse, since much of its theory and practice seemed to lead to nihilism, and nihilism was scarcely an advance over determinism!

CHARACTERISTICS OF THE PRESENT STAGE OF GEOGRAPHICAL THOUGHT

Thus, we come into the present period from about 1940 or 1945, or even later, which is difficult to define because we apparently are in the midst of it. As I interpret the written and unwritten record today and since the war, however, there appear to be six characteristics of the present (and future?) period differentiating it from the earlier ones. These tendencies do not eliminate the earlier periods which are far from watertight compartments, separated from the present, nor are they universally accepted. My version of the six characteristics of modern geography, particularly in America and Sweden follow, in no particular order, but arranged somewhat chronologically:

1. Increased emphasis on the topical or systematic—delving into a topical subject such as transportation or manufacturing or agriculture, rather than trying to cover everything. This allows for deeper penetration. This emphasis was given particular publicity by Edward Ackerman's publication in the *Annals of the Association of American Geographers* at the end of the war in which he recounted the lessons of wartime experiences of geographers.[2] Many geographers, paradoxically during the war argued for a systematic or topical divi-

"It is true that geography is what geographers do but this does not necessarily make it good. In fact if each new generation is not better than the old we are not developing the subject. I am not against the 'constant development of thought resulting from the reflective study of previous writing,' but I am afraid that to some people this appears to rule out originality and indicates that if something different is proposed it is automatically no good. While sympathetic to the traditions of the past, I feel it my duty to resist the temptations to rest on this tradition. The rather low reputation of geography among thoughtful people, whether correct or erroneous, encourages me all the more to try to look for something new.

What we need in geography is a conceptual approach, tempered by fact and observation, but not mere description. Likewise I sometimes wonder whether geography has been defined in terms of its most distinctive contribution by simply talking about the work of the past." [Comments on Richard Hartshorne's "What Kind of Science Is Geography?" (later published in *The Nature of Geography*)]

1. Richard Hartshorne, *The Nature of Geography* (Association of American Geographers, 1937), and the later, shorter volume, *Perspectives on the Nature of Geography* (Chicago: Rand McNally for Association of American Geographers, 1959).
2. Edward A. Ackerman, "Geographic Training, War Time Research and Immediate Professional Objectives," *Annals of the Association of American Geographers* 35 (1945):121–43.

sion of knowledge rather than a regional one, in contrast to many of the other social scientists.

2. A second characteristic of the present period, but also inherited from earlier periods, and one that bothers me somewhat therefore to put in, is the genetic or historical approach. In my terms, this is not necessarily part of the mainstream to be considered here, although it is a mainstream in its own right, with many workers.[3] One might equate it partially with the four characteristics listed below, however, by noting that the genetic approach does often involve problem solving, process, and diffusion or interaction.

3. A third, and very important, substantive characteristic of the present period might be characterized as the functional or interaction approach. This includes circulation, interaction, transportation, communication—a geography of movement, as opposed to a static geography. Much work is being done in this field in various forms. Transportation is peculiarly a geographic subject. It is the device for overcoming space, and space is of the essence in geography. I recall some years ago Professor Harold McCarty of the University of Iowa and Professor Richard Hartshorne of the University of Wisconsin presenting views at the Santa Monica meeting of the Association of American Geographers. McCarty spoke of a geography of production; Hartshorne, quoting Allix André, of the University of Lyon, France, of a geography of consumption. Afterwards, it came to me that really much better than either of their suggestions would have been a geography of circulation or transportation. The connections between production and consumption are peculiarly geographical.[4]

We don't, of course, judge the merits of an approach on whether it is peculiarly geographical or not. What we are interested in is whether or not these new trends can produce an approach which pays off. However, we are interested in whether or not the approach justifies geography as a separate, but related discipline.

4. An increased, in fact, new emphasis on quantitative approaches. By this, I mean measurement, both inductive and deductive, and the use of statistics and mathematical models. This is a characteristic of all sciences, and a characteristic of our age. Geography is also vigorously, if not always fruitfully, pursuing this method; it is important, in order to predict, to be able to put data into quantitative terms for manipulation. I will not assert, however, that the quantitative approach is everything. It is no substitute for ideas, concepts or empirical data. It is, in any case, a tool and properly used

3. Carl O. Sauer, *Land and Life: A Selection from the Writings of Carl Ortwin Sauer*, edited by John Leighley (Berkeley: University of California Press, 1963), and many other studies.

4. Edward L. Ullman, "The Role of Transportation and the Bases for Interaction," in *Man's Role in Changing the Face of the Earth*, ed. William L. Thomas (Chicago: University of Chicago Press, 1956), pp. 862–80; and *American Commodity Flow* (Seattle: University of Washington Press, 1957).

helps to produce substantive findings and theory. This brings us then to the fifth characteristic of the present period.

5. An emphasis on theory, which might be broken down into simple concepts and into systems of general order. These last three characteristics: the interaction approach, the quantitative approach, and the theoretical approach often are tied together.[5]

6. Finally, a sixth point, which feeds out of the preceding, would be application and prediction, the subject of the remainder of this presentation which will also furnish an example of the use of the five principal trends just noted: topical, functional or interaction, quantitative, theoretical, and predictive, illustrated by the operation of an actual prediction model applied to a region.[6]

These trends can also be reflected in a new, partial definition of geography: one pointed somewhat toward geography as spatial interaction. The goal of geography is the codification of relations between objects—occupance units—in earth space. (Attention is focused on economic or social relations promoting tangible interchange as measured by flow of goods or people, trade areas, or regional groupings, but not excluding movement of ideas.)

The basic assumptions are: (1) that the closer things are in space the closer are their relations; or (2) that there are definite categories of spatial relations not following the proximity rule, but exhibiting degrees of interconnection in response to varying degrees of spatial complementarity and mobility. They fall into patterns of contiguous or non-contiguous diffusion, into corridors of interaction, into waves or other manifestations of varying degrees of intensity and order, and into relations between separated nuclei of differing hierarchical or other characteristics.

Codification of these spatial relations into tendencies for understanding or laws in order to predict is the goal of scientific geography. The imaginative application of these spatial procedures and theories to a range of practical and social problems is the justification for applied geography.[7]

The Contribution of a Geographer

The contribution of a geographer is his knowledge of the human geography of some places and of some subjects, and his use of certain tools and theories. A geographer is a professional traveler, both actual and armchair. Armchair is probably more important because this implies heavy reliance on interpretation of maps, correlated with library, statistical,

5. See especially the extensive writings of Torsten Hägerstrand, William L. Garrison, Brian J. L. Berry, William Bunge, Duane Marble, Michael Dacey, William Warntz and others for quantitative and theoretical approaches.

6. Still other trends may be emerging. For example, considerable evidence points to a newly developing omnibus sort of characteristic composed perhaps of two parts: (1) a perception-behavior assessment often of irrational activities affecting space, and (2) an artistic-amenity appreciation of landscape. The first trend may well reflect a heightened psychological-social concern in America, and the second a rising concern with esthetics and amenity, in an increasingly affluent and educated society that values both urban design and natural beauty. These approaches will not be discussed further, but the writings of Wolpert on the first, and Lowenthal and the magazine *Landscape* (edited by J. B. Jackson) on the second, are illustrative. These approaches, especially the second, also are reflected somewhat in the amenity goals of the Meramec Project. (For calling my attention to these and some other points, I am indebted to Harold Brodsky.)

7. This definition was sparked in part as a complementary reaction to Edward A. Ackerman, *The Science of Geography* (Washington, D.C.: National Academy of Sciences, 1965).

field, and interview data. Therefore a geographer has a mental map—not exact—in his head of the world and certain regions, or analogous regions specialized on certain subjects. In varying detail he knows the size of cities, the routes of trade, the centers of agriculture and manufacturing, and concentrates on them in a rather precise way. He knows where he is and senses relations to other places, because of the generalizing quality of the map to reduce reality to a scale where larger relations, however imperfectly, can be visualized. This knowledge definitely is not of the rote variety of unweighted, unrelated data like state capitals or freaks of nature. It is conceptually arranged in varying detail, and is, of course, never complete in all aspects. But it is this knowledge that provides the basis for his theories, questions, and insights. It prompts him to seek similarities and note anomalies in the distribution of man and his works and in the use of the earth by man.

An alternative approach would be to rely on abstract spatial or other models and large multivariate, statistical, computer procedures. Such methods may indeed turn up some new and better weighted measures and conclusions and are increasingly employed. However, up to now, we cannot say that these methods have replaced a general insight approach. Indeed by using them we may often mechanically miss significant arrangements, deal with less interesting intellectual questions, or become involved in technique for its own sake. How can one answer the question why, at least in part, if one does not recognize an anomaly or a somewhat causal pattern? That many "answers" can be only probabilistic and unspecific is recognized and reinforces the need for knowledge.

One method that might have many virtues would be to substitute, in part, a real region, drawn from a somewhat similar universe, as an analog, instead of abstract space, adjust it, and use it as a model for a new problem in a new place. At least if one could not isolate all the variables, there would be a fair chance that they were there anyhow!

The geographer's value is in direct proportion to his knowledge, tools, and concepts. Some of these are common to all scholars, the better the mind and knowledge, the better the scholar. Some may be associated with one discipline, but by no means exclusively, as change or development is with history, or optimization and efficiency concepts are with economics. All should be incorporated as appropriate. Thus substantive knowledge, coupled with inquiring concepts, can provide a base for enriched, deeper penetration. That geographers had relied too much on simple description, and too little on theory and tools in the past is probably true. That too much reliance on abstract spatial theory or complex statistical manipulation is somewhat sterile is also probably true.

Use of the comparative method, coupled with theory, quantitative analysis, and specialized knowledge of some systematic topics, provides the basis for the discipline of geography.

The Test of Significance in Geography

One of the most naive and persistent questions in the minds of students is what one should study in geography. What phenomena are important? How does one go about making a significant contribution? If this is not a persistent question, it should be. It is the main point to which the voluminous discussions of geographic methodology presumably have formed an introduction. Answering the question, therefore, represents the pressing next step in advancing geographic thought and practice.

No pat answer is possible, since the measure of a contribution is the reflection of the intelligence and ability of the performer, as well as his luck in stumbling on relevant data. The last point should not be emphasized too strongly, however, since it is amazing what data can be turned up if one knows what to look for. Some guidelines particularly applicable to geography, but many of them also to other subjects, can be laid down to aid the intelligent beginner; they may also clarify some issues for the practiced performer, since no subject is static, and geography, although it has made substantial advances, is in no position to rest on its laurels.

These guidelines can be grouped under the following headings which will be amplified in the course of the paper (some appear obvious, but unfortunately some widespread practices or characteristics of our field as a descriptive science indicate that they require stating):

1. Is the contribution something new, either in idea or facts, not merely a repetition of the obvious?
2. Have proper criteria been used in assessing the relative importance of phenomena for study?
   a. In regional and much general geography: has proper account been taken of the significance of the phenomenon to other phenomena on the earth's surface, to the degree of interrelation or "zusammenhang"?
   b. In specific problems: are the data relevant to the problem at hand?
3. Is the research correct? Are the data, for example,

sufficiently quantitative, where possible, and well mapped or represented, to be most meaningful?

4. Does the research answer a question? How much valid causal explanation or consequences can one produce? Does the study have some depth, and not represent mere description?

5. What organizing concepts provide an intellectual frame for the study? What simple conceptual schemes, intermediate, or general theories come out of the study, or does the study pertain to?

A little reflection indicates that all these points not only overlap one another but are required for any task, yet a good job does not necessarily involve complete coverage of all. The second part of point 5, simple conceptual schemes, theory, etc., may not be required at all in an explicit manner although obviously some thought or concept is required before even inductive work can be contemplated. Particularly desirable, therefore, is the first part of point 5, some organizing concepts, etc., to serve as a frame, even if it be only at the level of such simple, familiar concepts as site or situation. The whole of point 5 will, however, be emphasized as the most specific way of gaining significance.

Omitted from this list is a third part of point 2, namely, to select phenomena that relate to the earth as a home for man, not for ants, or fish, or what have you. This is omitted because this paper by definition will be limited to human or cultural geography, broadly conceived, not physical geography.

Emphasis throughout will be on a distinctively geographical contribution, although there is no objection to chasing a problem wherever it leads. In fact this procedure is one of the major tests of the significance of a contribution; the study should be related to other fields as much as possible. In some cases the study may take one out of the field of geography altogether. The contribution, if done correctly, however, remains a contribution; it is merely in another field. Our concern, to repeat, however, will be to stress distinctively geographical ways of thinking and arriving at fruitful conclusions, if for no other reason than to demonstrate that there is a rich field of our own needing cultivation. All possibilities, even within geography, however, will not be covered.

Omitted also from this discussion, therefore, will be any attempt to fit geography into any standard scientific strait jacket starting with axioms and theory and winding up with rigorous theoretical or actual proof. These scientific procedures have been discussed voluminously and persuasively by others; I do not wish to repeat them and thus violate my first principle any more than I have to. (Besides I always forget the wording of these elegant standard hortatory introductory tracts, a weakness of character to which I humbly confess.) By the same token I shall not address myself directly to many praiseworthy objectives, such as precise prediction in geography, the goal of science, according to many, or to optimization, maximization, or efficiency techniques as practiced in economics, other social sciences, or operations research, and particularly to the question of solving a practical problem or even directly clarifying a public issue. The concern at this stage is with understanding, which also involves theory, and not practical solutions; by analogy, for example, it seems more significant to know that there is a solar system with planets revolving around the sun than to be able to predict the tides from the moon's phases, however more practically useful the latter may be.

"Perhaps the essential intellectual contribution of human geography can be summarized by the concepts of site and situation. Site refers to local, underlying areal conditions and leads to defining geography as the study of the 'relations between man and the environment.' Situation refers to the effects of one area, or rather phenomena in one area, on another area. . . ." ["Geography as Spatial Interaction"]

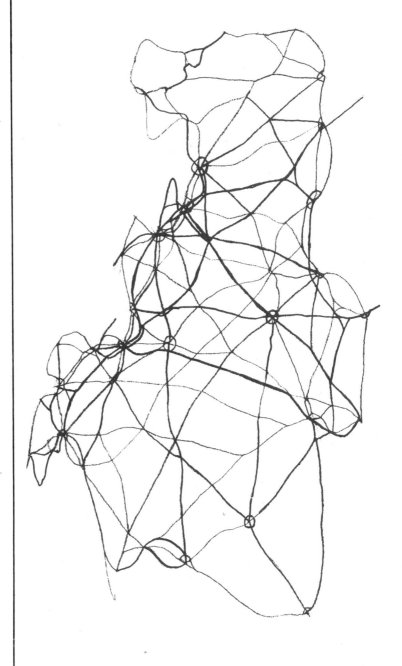

CHAPTER 1   Geography as Spatial Interaction

"Geography as Spatial Interaction" *appeared in 1954 in a rather obscure publication,* Interregional Linkages, *the Proceedings of the Western Committee on Regional Economic Analysis (Berkeley, California), pp. 63–71. The paper here presented is a larger draft version that, to my knowledge, has never been published before. Ullman gave this paper in abbreviated fashion at the Association of American Geographers' meeting in Philadelphia on 12 April 1954. Although he did not follow the article through to national publication, he presented many of the ideas in a chapter, coauthored with Harold Mayer, "Transportation Geography," in* American Geography: Inventory and Prospect *(Syracuse University Press, 1954), pp. 310–32.*

Perhaps the essential intellectual contribution of human geography can be summarized by the concepts of site and situation. Site refers to local, underlying areal conditions and leads to defining geography as the study of the "relations between man and the environment." Situation refers to the effects of one area, or rather phenomena in one area, on another area. It should logically focus on the connections between areas and leads to such terms as "circulation" and "regional interdependence" or to specific aspects such as "diffusion" or "centralization."

This situational concept is defined here as "spatial interaction," and is intended as a more positive and dynamic concept than either situation, relations, or even circulation. In a sense it provides a "motor" for situation concepts.

Site thus might be conceived of as a *vertical* relationship—type of soil correlated with type of agriculture on top of the soil—and situation a *horizontal* relationship—effect of market in one place on type of agriculture in some other area. As early as 1889 Sir Halford Mackinder noted the same dualism as follows: "The chief distinction in political geography seems to be founded on the facts that man travels and man settles."[1]

An example of alternate interpretation based on site and situation is provided by the age-old puzzle of assigning reasons for the growth of particular civilizations in particular places. Thus Toynbee in his challenge and response theory uses a site concept with a new twist—the challenging effect of a relatively poor environment. Pierre Gourou, the Belgian geographer, asks whether substitution of the effects of an unfavorable environment for a favorable one represents an advance over previous environmental determinism. He poses as an alternate possibility a situation concept—the rise of civilizations in favored corridors for interaction so that contact with other civilizations and contrasting ideas was facilitated, as in parts of Europe.[2] Without going into the merits of either explanation, undoubtedly site, situation, and other factors are all involved in any total understanding.

In practice the two concepts, site or relation of man to environment, and situation or spatial interaction, are intermingled in application, as Mackinder implied, and probably represent extremes on a continuum. It is important, however, to

1. H. J. Mackinder, "The Physical Basis of Political Geography," *Scottish Geographical Magazine* 6 (1890):78–84.

2. Pierre Gourou, "Civilisations et malchance géographique," *Annales, économies, sociétés, civilisations* (October–December 1949):445–50.

recognize the two, if for no other reason than that the first concept has received the greatest attention and the second has tended to be ignored. The late recognition and acceptance by geographers of the concept of a functional or nodal region is an illustration.[3] If the vertical or site type of study has preoccupied most geographical work, this does not mean it is either more important or more distinctly geographical. Thus Hartshorne approves Hettner's conclusion: "No phenomenon on the earth surface may be considered for itself; it is understandable only through the apprehension of its location with reference to other places on the earth."[4] Others have explicitly argued for the concept in recent years, although using different terms. In 1932 Whitaker published an eloquent brief paper on regional interdependence.[5] He applied the concept quite properly to both physical and human geography. Because of limitations of space this paper will deal only with human geography.

In 1949 Platt presented to the American Association of Geographers a lively argument for the concept and an example, subsequently published, based on Tierra del Fuego.[6] He also noted, optimistically, that geography had already gone beyond the concepts and practices emphasized by Hartshorne's pre-war publication, *The Nature of Geography*, which he said should be regarded as a milestone, not a tombstone. Platt also adds quite properly a historical, dynamic dimension to his functional treatment. Perhaps the strongest plea for a dynamic or kinetic approach, however, was in a paper by P. R. Crowe entitled "On Progress in Geography" in the *Scottish Geographical Magazine* for 1938.[7] Crowe severely and properly, I believe, criticized geography for not getting very far with a morphological, descriptive approach and suggested that a more fruitful study would be to concentrate on currents and men and things moving rather than on static description of facilities and areas.[8]

In view of the at least implicit recognition given the concept it is surprising how little work has been done with it. Hartshorne notes this and suggests that perhaps the reason why relative location, as he calls it, is largely ignored is because it is inconvenient in a regional treatment constantly to leave a region and move out of it.[9] One answer to this obviously is to focus on the interaction. This alone provides a good excuse for this paper.

The few works which do stress the interaction concept (except for Platt's regional example), unfortunately, give us few clues for making it operational—for measuring interaction, classifying the types, and attempting to erect a theory or sys-

". . . the study of transportation, by its very nature perhaps man's most fundamental dynamic areal expression, presents an intriguing problem. . . ." [Letter to Charles C. Colby, 15 June 1936]

3. Cf. G. W. S. Robinson, "The Geographical Region: Form and Function," *Scottish Geographical Magazine* 69 (1953):49–58, or D. S. Whittlesey, "The Regional Concept and the Regional Method" in *American Geography: Inventory and Prospect* ed. Preston E. James and Clarence F. Jones (Syracuse University Press for the Association of American Geographers, 1954), pp. 36–44.

4. Richard Hartshorne, *The Nature of Geography* (Lancaster, Pa.: Association of American Geographers, 1939), p. 283.

5. J. R. Whitaker, "Regional Interdependence," *Journal of Geography* 31 (1932):164–65.

6. R. S. Platt, "Reconnaissance in Dynamic Regional Geography: Tierra del Fuego," *Revista geografica* (Rio de Janeiro) 5–8 (1949):3–22.

7. P. R. Crowe, "On Progress in Geography," *Scottish Geographical Magazine* 54 (1938):1–19.

8. Elsewhere in the world appreciation of features of the concept is found. Cf. Aldo Sestini, "L'organizzazione umana dello spazio terrestre," *Revista geografica italiana* 59 (1952):73–92.

9. Richard Hartshorne, *The Nature of Geography*.

"I incline, at the moment, to define human geography and all related fields as 'the study of spatial relations,' especially in relation to man and his work." [Letter to William L. Thomas, Jr., 20 June 1951]

"After I have read my paper at Philadelphia on 'Geography as Spatial Interaction,' I hope that my argument will be so persuasive that a lot of good people will enter transportation geography." [Letter to Derwent Whittlesey, 1 April 1954]

tem of explanation. This paper will suggest a system, based on previous examination of new quantitative measures of commodity and other flows. The complexities inherent in the many and different types of interaction will also be noted along with some errors in previous particularistic explanations based on erroneous single factor analysis.

THE BASES FOR INTERACTION

Complementarity

It has been asserted that circulation or interaction is a result of areal differentiation; to a degree this is true but mere differentiation does not produce interchange. Numerous different areas in the world have no connections with each other.

In order to have interaction between two areas there must be a demand in one and a supply in the other. Thus an automobile industry in one area would use the tires produced in another but not the buggy whips produced in still another. Specific complementarity is required before interchange takes place. Complementarity is thus the first factor in an interaction system; because it makes possible the establishment of transport routes.

So important is complementarity that relatively low-value bulk products move all over the world, utilizing, it is true, relatively cheap water transport for most of the haul. Some cheap products in the distant interior of continents, however, also move long distances. Thus when the steel mills were built in Chicago they required coking coal and reached out as far as West Virginia to get suitable supplies, in spite of the fact that the distance was more than five hundred miles by land transport and the coal was relatively low value.

Complementarity is a function both of natural and cultural areal differentiation and of areal differentiation based simply on the operation of economies of scale.[10] One larger plant may be so much more economical than several smaller ones that it can afford to import raw materials and ship finished products great distances, and thus interaction may take place between two apparently similar regions, such as shipment of specialized logging equipment from Washington to forest areas of the South. In this case the similarity in other respects of the two regions provides the market and encourages the interaction. This, however, is insufficient to affect significantly many total interactions because specialized products dominate the total trade of many regions. Thus total

10. Cf. Bertil Ohlin, *Interregional and International Trade* (Cambridge, Mass.: Harvard University Press, 1933).

shipments from Washington to southern states are low because of the dominance of forest products in each (figs. 1.1 and 1.2).

Another example of similarity producing complementarity is provided by the overseas Chinese who furnish a significant market for the mother country's export handicrafts and other products.[11] The same occurs with Italians and other transplanted nationals. Perhaps one could generalize and say that similar cultures but different natural environments tend to promote interchange.

Intervening opportunity

Complementarity, however, generates interchange between two areas only if no intervening, complementary source of supply is available. Thus few forest products moved from the Pacific Northwest to the markets of the interior northeast sixty years ago, primarily because the Great Lakes area provided an intervening source. Florida attracts more amenity migrants from the Northeast than does more distant California. Many fewer people probably go from New Haven to Philadelphia than would be the case if there were no New York City in-between as an intervening opportunity. This, presumably, is a manifestation of Stouffer's law of intervening opportunity,[12] a fundamental determinant of spatial interaction, and the second factor proposed for a system of explanation.

Under certain circumstances intervening opportunity might ultimately help to create interaction between distant complementary areas by providing a nearby complementary source which would make construction of transport routes profitable and thus pay for part of the cost of constructing a route to the more distant source. On a small scale this process is followed in building logging railroads; the line is extended bit by bit as timber is cut nearer the mill and ultimately trains are run long distances between mill and supply as the nearby supplies are exhausted. If the line had to be constructed the long distance initially, it might never have been built. On a larger and more complex scale this is what happens in transcontinental railroads—every effort is made to develop way business, and as this business develops it contributes to some of the fixed costs for long distance interchange.

Distance

A final factor required in an interaction system is distance, measured in real terms of time and cost. If the distance

"Early on I became interested in two things—measuring interactions and the role of distance. Theory came along, I don't know from where, perhaps abetted by an economist, but I guess understanding interaction and distance (which were largely original when I started) remain my principal concern, but theory (especially grand theory) is less promising. Probability is something else." [Letter to Michael Dacey, 9 February 1976]

11. Theodore Herman, *An Analysis of China's Export Handicraft Industries to 1930* (Ph.D. dissertation, Department of Geography, University of Washington, Seattle, 1954).
12. S. Stouffer, "Intervening Opportunities: A Theory Relating Mobility to Distances," *American Sociological Review* 15 (1940):845–67. (This theory was applied to intra-city migration, but the concept, with appropriate modification of detail is applicable to other interactions, I believe.)

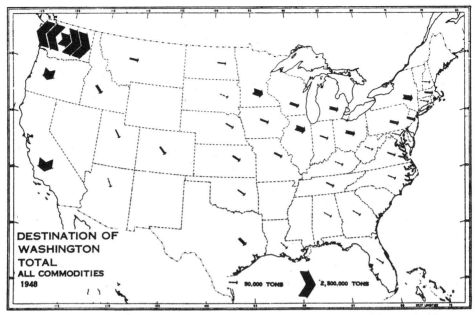

Fig. 1.1. Destination, by states, of total commodities shipped by rail from Washington, 1948. (Tons are short tons of 2,000 pounds.)

Fig. 1.2. Destination, by states, of forest products shipped by rail from Washington, 1948. Width of arrows is proportionate to volume; arrows within Washington represent intra-state movements. (Tons are short tons of 2,000 pounds.)

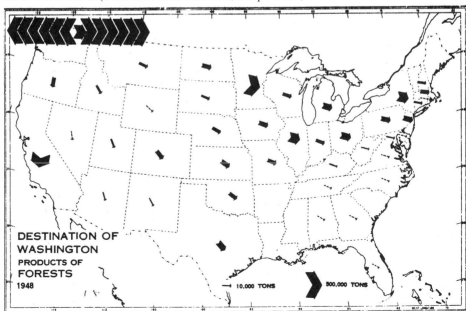

between market and supply were too great and too costly to overcome, interaction would not take place in spite of perfect complementarity and lack of intervening opportunity. Thus alternate goods would be substituted where possible; bricks would be used instead of wood, etc.

Thus we might consider that the factor of *intervening opportunity* results in a *substitution of areas* and the factor of *distance* results in a *substitution of products*.

It is a mistake, therefore, to assume that all places, even giant commercial centers, are linked equally with other producing areas and centers of the world. Distance and intervening opportunity drastically trim down the relative quantity of such dramatic, long-distance relationships which international trade enthusiasts like to emphasize. Great Britain and the United States provide two contrasting examples. To reach enough complementary sources Britain must trade with the world. The United States, on the other hand, can reach sufficiently complementary areas merely by trading within its own borders to account for the overwhelming bulk of its trade, with much of the remainder coming from Canada and the nearby Caribbean.

To sum up: a system explaining material interaction can be based on three factors: (1) *complementarity*—a function of areal differentiation promoting spatial interaction; (2) *intervening complementarity* (or "opportunities") between two regions or places; (3) *distance*, measured in real terms including cost and time of transport and effect of improvement in facilities.

The system proposed applies primarily to interaction based on physical movement, principally of goods, but also to a large extent of people. It does not apply to the spread of ideas or most other types of communication, except as they accompany the flow of goods or people, which admittedly is often the case. Intervening opportunity, for example, would seem to facilitate rather than check the spread of ideas. Similarity of two regions also probably would facilitate the spread of ideas more than difference or complementarity, although the latter would be important for some cases.

An empirical formula often employed to describe many types of interaction is a gravity model which states that interaction between two places is directly proportionate to the product of the populations or some other measures of volume of two places and inversely proportionate to the distance (or distance to some exponent) apart of the two areas. This measure is often written $P_1 P_2 / d$ (population place[1] times population place[2] divided by d, the distance apart of the two places).

This model, however, is useless in describing many interactions because it assumes perfect or near perfect complementarity, a condition that seldom obtains for physical flows. Some form of the model apparently does come close to describing many interchanges, even for goods in a few cases, but apparently primarily for many more or less universal, undifferentiated types of flow such as migration of some people or telephone calls between cities. It has been developed by Zipf, Stewart, Dodd, and others.[13]

The three-factor system of complementarity, intervening opportunity, and distance, however, I believe, will cover any case of material interaction of goods or people. The system should be kept in mind by investigators lest they be led astray by assigning exclusive weight to only one of the factors in attempting to explain past interactions or in predicting potential interaction when underlying conditions change.

Traffic vs. Facilities as Generators of Interchange

Examples of erroneous single-factor explanations are numerous. One type concerns the relative role of traffic vs. facilities as promoters of interchange. Thus New York City was the largest port in the United States before the Erie Canal was built, as Albion has shown,[14] and its size, plus some settlement in the West made feasible construction of the Erie Canal just as the opening of the Canal had the effect later of drastically cutting real distance and thus enormously facilitating interchange and the growth of New York. Likewise, the great voyages of discovery were made in large part to tap the growing traffic between the Orient and Europe. Between these two centers were no significant intervening opportunities, although some were discovered as the routes were developed.

A more detailed example is provided by the opening of the St. Gotthard Pass across the Swiss Alps in the thirteenth century.[15] According to an earlier, ingenious interpretation by the German historian, Aloys Schulte, in 1900, it was the invention and construction of a suspension, chain bridge along the vertical walls of the Gorges of Schoellenen that opened up this best of all passes and produced a flood of traffic through Switzerland. Thus it was not William Tell who founded the independence of Switzerland but an unknown blacksmith who built the chain bridge that opened Switzerland to the currents of freedom from the south and the trade to support many people. Twenty-five years later careful research by scholars changed this interpretation indicating: (1)

13. Cf. G. K. Zipf, *Human Behavior and the Principle of Least Effort* (Cambridge, Mass., 1949); J. Q. Stewart, "Empirical Mathematical Rules Concerning the Distribution and Equilibrium of Population," *Geographical Review* 37 (1947):461–85; S. C. Dodd, "The Interactance Hypotheses: A Gravity Model Fitting Physical Masses and Human Behavior," *American Sociological Review* 15 (1950):245–56; Joseph A. Cavanaugh, "Formulation, Analysis and Testing of the Interactance Hypotheses," *American Sociological Review* 15 (1950):763–66; and other writings by the same and other authors. It should be noted that interaction in sociology is defined more narrowly than its use in this paper.

14. R. Albion, *Rise of New York Port* (New York: C. Scribner's Sons, 1939).

15. Charles Gilliard, "L'ouverture du Gothard," *Annales d'historie economique et sociale* 1 (1929):177–82.

that before the hanging bridge was built, the precipitous
gorge was actually by-passed without too great difficulty by
taking a longer route through Oberalp; (2) hanging bridges of
the type noted were in reality common in the Alps by the
thirteenth century; (3) the key bridge was not really the one
indicated but rather another farther down, which was built of
stone masonry by an unknown mason, requiring much more
effort and capital than a mere suspension bridge; and, finally
(4) (and most important), this key bridge and the rest of the
route were not built until traffic was sufficient to pay for
them! The traffic was generated by increased activity in the
complementary regions of Flanders and the upper Po Valley,
between which were few intervening complementary
sources. Thus one must conclude that traffic was equally, if
not more, instrumental in creating the route than was provision of the route.

A still different type of erroneous single-factor analysis
concerns the role of certain features of the natural environment in promoting or retarding interchange. Mountain
ranges for example are commonly thought of as barriers to
interchange but in many cases their barrier quality may be
more than compensated for by the differentiation or complementarity which they produce. Thus climate, in many instances, differs on two sides of a mountain range; this difference may create interchange. More directly the mountains
themselves may be so different as to generate interaction, as
in the case of transhumance—the moving of animals from
lowland winter pastures to mountain summer pastures. Even
more important in the modern world is the production of
minerals in mountains associated with folding, faulting, un-
covering of subsurface deposits by stream erosion, or other
factors. The central Appalachians thus provide enormous
quantities of coal producing the largest single commodity
flow in America. The Colorado Rockies, because of minerals,
at one time had a denser network of rail lines than neighboring plains areas, in spite of formidable difficulties of penetration.

Potential New Interaction

An example of the second reason for using the system, to
predict or understand potential new interaction under
changed conditions, is provided by Portland, Maine, and
Canada. At the end of the nineteenth century Portland was
known as the winter outlet for Canada because it was the
nearest ice-free port. The Grand-Trunk Railroad built a line to

the city and extensive docks. Canadian wheat was shipped out in quantity. Then Canada decided to keep the wheat flows within its borders and diverted the trade to the more distant ice-free, ocean ports of St. Johns and Halifax in the Maritime Provinces of Canada. Portland declined. Recently two changes have occurred. During World War II a pipeline for gasoline was constructed from Portland to Montreal to save long tanker trips through submarine-infested waters and to insure a year-around supply. This gave Portland a shot in the arm and resulted in construction of large tank farms.

The second change can be illustrated by a story. In the summer of 1950, on a Sunday night, I stood on the international border between Derby Line, Vermont, and Rock Island, Quebec, and marveled at the constant stream of cars returning to Canada. I asked the customs inspector the reason, and he replied, "Ninety percent of the cars are bound for Quebec City and are coming from Old Orchard Beach, Maine." Old Orchard Beach is near Portland and is the nearest ocean beach to parts of eastern Canada, just as Portland is the nearest ocean port. The Dominion government in this case could hardly force tourists to drive a whole extra day to reach the Maritimes. Thus (1) Portland's potential complementarity reasserted itself, (2) no intervening opportunity (ocean beach) occurs between Portland and Quebec, and (3) the distance is short enough so that it can be driven in a long week-end trip. Presumably if the distance were much greater, residents of Quebec would confine their swimming to the bath tub and use sun lamps. Needless to say, the underlying change permitting both interactions was the invention and development of the automobile, and, in conjunction with the tourist movement, increased leisure and higher standard of living, both fundamental trends, especially in Anglo-America!

A similar example, but one in the nature of a prediction, is the reasonable expectation by Professor Folke Kristensson, of the Stockholm School of Economics, that as living standards rise in Sweden, Swedish diet will change, as the American diet has, and more year-round fresh fruits and vegetables will be consumed. This will result in increased interaction between Sweden and the nearest complementary sources— Italy, Southern France, North Africa, etc., just as occurred between the northeastern United States, Florida, and California, a fundamental feature today of the American interaction pattern. In fact, it is difficult to conceive of any changes, technical, political, social, or economic, which do not have some effect on interaction patterns.

FURTHER CHARACTERISTICS,
CONDITIONS, AND EFFECTS OF INTERACTION

Once the decision is made to focus on interaction, a host of topics and subconcepts present themselves for analysis. Space allows only sample treatment of the following: (a) two characteristics of interaction, direction and length of haul; (b) some conditioning factors, especially type of commodity and political sovereignty; and (c) the effect of increased interaction on areal differentiation.

Direction

Movement along a route is generally unbalanced, a probable result of complementarity because it is unlikely that each of two regions will have exactly what the other wants in like quantity. However, strong forces work to promote a balance for two reasons: (1) to make transportation of many goods and persons more economical by using facilities fully, and (2) to pay for the import.

The second factor may be taken care of by return flows of different types—capital movements, tourist or immigrant remittances, triangular or multilateral trade, gradual shift of population, etc.

The first type, to balance movement in order to make transportation cheaper, often results in establishing low return haul or "ballast" rates. The classical example of this was the export of coal from England to counterbalance imports of bulky raw materials. The coal shipments reached such great proportions that they actually unbalanced the trade in the other direction in the early twentieth century.

In other cases a return load may not be obtainable, but compensation is provided in other ways. Many United States transcontinental railroads, for example, have rebuilt their lines so that their ruling grades eastbound, the direction of heavy flows representing heavy western raw materials moving to the Eastern market, are less than the ruling grades westbound, the light direction.

In ocean shipping, and even land transport, much triangular (or multi-angular!) trade is carried on in order to obtain full loads.

In many areas, however, even this is not feasible as in the Olympic Peninsula of Washington where heavy raw materials are shipped out by rail and ship and finished products are brought in by truck. This produces high transport costs per mile.

For many media of interaction the problem does not arise. Pipelines run even better on one-way flows, part of the reason for the low cost of this form of transportation, which already covers more route miles than railroads in the United States, and one of the reasons why industries and population do not move to oil as much as to coal. Communication facilities (radio, telephone, telegraph) work in similar fashion. In fact radio and television provide spectacular examples of overwhelmingly greater cheapness for one-way flow, as witness the few transmitters and the numerous receivers. Nor should it be forgotten that large volume in one direction can itself produce low rates. Coal is an example; tankers, outside of a very few trades, also apparently find it unprofitable either to alter their construction to carry other cargo or to move to other wharves in order to obtain return hauls.

Length of Haul and Break of Bulk

It is a general rule that the longer the haul the less the cost per mile. Thus the effect of distance is not directly proportional to increases. Distant areas characteristically are blanketed in freight rates, the nearer distant point having the same rate as the more distant one. Thus steamship rates from the United States generally are the same to any port in the Havre-Hamburg range. Related to this is the high cost of breaking bulk—unloading or loading, and terminal charges. For both these reasons end points along a flow line, either at raw material or market or at one raw material when two are used, are favored locations for processing, as Hoover and others have noted.[16] This is probably an underlying factor in the higher freight rates on ore and coal combined, which the steel industry in the Valley district around Youngstown pays over Cleveland or Pittsburgh, the sources, or break of bulk points, for iron ore and coal respectively.[17]

The lower cost per mile of long hauls applies to all forms of transport, but particularly to ocean trade where terminal costs are high. Nevertheless, slight increases in distance on long hauls may be reflected in subsequent development. Thus Liverpool and west of England ports concentrate on the Atlantic trade and Eastern ports on the North Sea and Baltic trades.

Under certain special conditions the underlying influence of short distance increases on long hauls is given free play to operate. During World War II American troops were billeted in the west of the invasion of Europe in order to avoid entangling U.S. supply lines with the British. The final result of

16. E. M. Hoover, *The Location of Economic Activity* (New York: McGraw-Hill, 1948).

17. Cf. Allan Rodgers, "The Iron and Steel Industry of the Mahoning and Shenango Valleys," *Economic Geography* 28 (1952): 331–42.

the U.S. armies moving across Europe on the right flank, according to Lt. Gen. Sir Frederick Morgan, was the establishment of the United States zone in Germany in the south and the British zone in the north.[18] Thus is seen in operation a peculiar (or "parallel") long distance effect of position but one in accord with logical reasons, once the conditions prevailing are known. Somewhat similar parallel corridors are provided by U.S. transcontinental rail routes which connect Seattle with St. Paul or Los Angeles with Kansas City.

Effect of Type Commodity on Movement

It appears reasonable that low value goods would tend to move shorter distances than high value commodities. To a large degree this is true; thus gold is mined in the mountains of Central New Guinea and flown out; bricks, sand and gravel are rarely shipped long distances. Low value commodities are also charged lower rates per mile to encourage their shipment presumably because they cannot bear the higher costs.[19] Many exceptions exist to this rule however. If no alternate sources (intervening opportunities) of supply for necessities exist they may move relatively long distances even over land as witness some historic salt trades or even the flow of Pocahontas coal as far east as New England, south to Georgia and west to North Dakota. Furthermore low value goods that can be handled in bulk can be loaded and unloaded far more cheaply than higher value packaged goods, and thus take advantage of water transportation. Any commodity, to use the Maritime Commission terms, that can be "scooped, dumped, poured, pumped, blown or sucked" has great mobility in many areas. Oil also is piped overland across half of America. And finally, as is well known, transport costs over the oceans are relatively so low that many ores or grain can move around the world by water. Still other types of flows are differentially affected by barriers. High tension lines and natural gas pipe lines, for example, can convey energy across mountains relatively easily and cheaply.

Political Sovereignty as a Conditioning Factor

Few influences have been greater in distorting spatial interaction than political control of areas and the consequent channeling of trade and interaction. This is so well known as to need little amplification. However to drive home the point, two extreme examples may be noted: (1) the enormous reach of the Russian Empire in the nineteenth century following the

18. Sir Frederick Morgan, *Overture to Overlord* (Garden City, N.Y.: Doubleday, 1950), pp. 113–14 and 247–50.

19. Cf. E. F. Penrose, "The Place of Transport in Economic and Political Geography," *Transport and Communications Review* 5 (1952):1–8.

fur trade across Siberia to Alaska and even down the Pacific Coast as far south as California; (2) the control by Spain of America from California on the North to Argentina on the South from the seventeenth to the early nineteenth centuries. Trade with Buenos Aires was allowed only via routes running across the Atlantic from Spain to Panama and then south along the length of the Andes. No wonder that the British found it profitable to smuggle directly into Buenos Aires from the sea!

In both these cases the interaction evolved step by step in a sort of domino fashion until a literally non-viable arrangement was produced, which could not be economically maintained.

Two Other Conditioning Circumstances

Examples of two other conditioning circumstances which time precludes treating are:

1. Temporal characteristics, particularly war conditions which will produce quite different interactions from peace time, as use of air power indicates.

2. Types of milieu within which interaction occurs: intraurban areas with heavy passenger flows, and extraurban areas with heavy commodity flows; ocean realms in which some cultures possess sea skills—as in Greece, England, or Polynesia—and continental realms where some cultures exhibited great overland mobility—as the Mongols or Kazaks did for horsemen but not for goods.

Effects of Increased Interaction on Areal Differentiation

A feature of the modern commercial world is the enormously increased flow of goods and peoples facilitated by revolutionary improvements in transportation and communication. The great trade routes of the past were mere trickles compared to today's volume movements.

It is often asserted that this increased transportation and trade has accentuated areal differentiation by allowing areas to specialize on what they can do best. To a great degree this is true, but the more significant effect has been to change the *scale* of areal differentiation. Cheap transportation enables large areas such as the corn belt or wheat belts to specialize; thus they are more sharply differentiated from their neighbors, but within these areas there is less differentiation. On a small scale therefore increased interaction has often tended to produce greater uniformity. This contrasts with conditions prior to modern transportation as shown by von Thünen's famous model of the *Isolierte Staat* with concentric rings of different land uses around a city arising in response to cost of transportation to market. Now transport costs are relatively so low in accessible areas that other characteristics of production and rent-paying ability are generally more controlling than transport costs.

A second common notion is that the transportation pattern and rate structure produce an artificial geography which presumably prevents the best sort of areal specialization. To a degree this may be true, but I suggest that a more correct generalization may be that freight rates and provision of transportation tend to accentuate and perpetuate initial areal differentiation. Freight rates in the United States are generally made on a commodity basis; low rates are granted on volume movements which specialization tends to provide. Thus new areas or small producers may find it difficult to compete initially. John Alexander found that the fertile cash grain area of Central Illinois had low rates per mile to principal markets for its main product, corn, but high rates on cattle, whereas rates in the less fertile cattle-producing area of Western Illinois were reversed—low on cattle and high on corn.[20] The rate structure thus tended to accentuate and perpetuate areal specialization based on natural conditions.

CONCLUSION

Site and situation concepts and their extension to cover the relation of man to environment and to spatial interaction furnish in broad fashion perhaps the main basis for geographical theory.

This paper has attempted to provide the beginnings of a system explaining the basis of spatial interaction, a system based on complementarity, intervening opportunity, and distance. Perhaps route might be a fourth factor, although it is largely subsumed under distance. Additional generalizations and hypotheses covering important subsystems or topics have also been attempted. These hypotheses, in large part, have grown out of consideration of many new quantitative data, principally mapping of traffic flow and origins and destinations, which space has prevented presenting or discussing in detail. Undoubtedly many of the generalizations can be refined and some may be superseded by others. Still other new ones await discovery and development.

Other disciplines also are increasingly concerning themselves with interaction, although different labels may be at-

20. John W. Alexander, "Freight Rates as a Geographic Factor in Illinois," *Economic Geography* 20 (1944):25–30.

tached to it. In economics the term linkages is used, and appears to be a topic of growing interest, especially among regional economists.[21] In international relations, in the field of political science, study of interaction patterns has been called "one of the two basic ways to describe and explain international politics," the other being a decision-making approach.[22] In history and other fields the diffusion of ideas and their effects has been treated often, and is considered a major unifying thesis by some.[23]

In sociology, interaction is extensively investigated although often defined somewhat more narrowly and specifically than in this paper. Some sociologists even define sociology as the study of social interaction; by the same token geography might be defined as spatial interaction.

The purpose of this paper therefore has been to try to make explicit that which has only been implicit in most geographical writing. The concept, like everything else, is not new. Focusing research on interaction may well provide a fruitful avenue of advance for many disciplines.

21. As witness the program of the third annual meeting of the SSRC Western Committee on Regional Economic Analysis, Berkeley, 24–26 June 1954.

22. Karl W. Deutsch, *Political Community at the International Level: Problems of Definition and Measurement*. Introduction by Richard C. Snyder. (Foreign Policy Analysis Series No. 2, Princeton University, September 1953.)

23. Cf. Gilbert Highet, *The Migration of Ideas* (New York: Oxford University Press, 1954).

Human Geography and Area Research   CHAPTER 2

At first glance it might appear that only in physical geography would a geographer make a contribution to an area research program staffed by social scientists.* Geographers, however, do not so delimit their field.[1] This paper focuses in particular on the contributions of human geography. In the treatment I present my own point of view, but am only too conscious that I neither reveal exclusively new thoughts nor, in the attempt to emphasize new ideas, cover all topics uniformly.[2]

As an example of the distinctive nature of human geography let us consider economic geography. This branch, Wooldridge and East observe, is in practice at least three quarters of the field of human geography and (rightly or wrongly) is not a branch of economics any more than is economic geology.[3] Hartshorne, Finch, and others earlier indicated much the same opinion.[4] Most topics studied by the economic geographer are also considered by economists in various specialist branches of that subject, but are not, however, core topics in economics. Some of the topics which the economic geographer studies include various aspects of agriculture, mining, and other production; industrial location and development; transport routes and flows; markets; and rural and urban settlements. More nearly core topics for a relatively "pure" economist, on the other hand, include price movements, business cycles, competition, personal and national income, taxation, finance, and fiscal policy generally.[5]

Most conventional economic theory has little relevance to geography,[6] although there is room for development of a more "economic" economic geography,[7] along with still other growth stemming from the concepts of geography itself and other disciplines. Nor is economic geography particularly a part of technology. We have no more interest in the production process inside a factory than has an economic theorist. What we are mainly interested in, as Professor Tower noted years ago, is what comes in the back door and what goes out the front door of a plant, so that we may know why it is located where it is and what its effect on the area under study is. The effect of changes in internal processes, however, are relevant in so far as they affect these external relations.

*Revision of talk given to Interdisciplinary Area Studies Symposium, Association of American Geographers, Washington, D.C., 7 August 1952. Thanks are due the Office of Naval Research for support of research for parts of this paper.

This article was first published in the Annals of the Association of American Geographers *in March 1953 (pp. 54–66). In many respects it was meant to be a geographer's definition of geography. Although a believer in an interdisciplinary approach to problems, Ullman considered himself "primarily a geographer's geographer, with the many*

1. In this connection it should be emphasized that in terms of either subject matter or approach the individual social scientist often finds himself more *en rapport* professionally with some of his colleagues in another discipline than with many in his own field. This reflects the specialties within, and the obvious limitations of, the boundaries of conventional disciplines. In addition each of the specialists in an area program might well call in other specialists and sub-specialists almost ad infinitum—hydrologists, botanists, psychologists, engineers, etc., although in practice the major specialist would dig up much of such material himself, since there is nothing to prevent him from reading, listening, and learning.

2. I hereby acknowledge my debt for ideas I have consciously and unconsciously borrowed (and twisted) from a company so numerous and well-known that I do not need to list them separately. This applies particularly to Richard Hartshorne, *The Nature of Geography*, 1951 and 1949; earlier editions: 1946 and 1939, including *Annals of the Association of American Geographers* 29:171–658.

3. S. W. Wooldridge and W. G. East, *The Spirit and Purpose of Geography* (London and New York: Hutchinson's University Library, 1951), p. 114.

4. V. C. Finch, "Training for Research in Economic Geography," *Annals of the Association of American Geographers* 34:207–15.

weaknesses that this implies" (letter to Richard Hartshorne, December 1955), whose task was to clarify and develop geography from within as an independent field. Ullman liked to say that geography is what geographers do. "I first heard this phrase from Wellington D. Jones in about 1938," he said. "I don't know how long he had been using it nor where he got it. As a purist Jones cautioned me to add: 'when acting professionally' " (letter to J. Wreford Watson, 15 September 1953). Ullman always qualified it further by saying, "when acting both professionally and geographically."

5. Adapted from R. J. Harrison Church, "The Case for Colonial Geography," *Transactions and Papers, 1948* (Institute of British Geographers, publication no. 14), pp. 21–22.

6. Cf. Isard's remark that economic theorists are chiefly concerned with introducing the time element into their analyses, and note his quotation from Marshall: "The difficulties of the problem depend chiefly on variations in the area of space, and the period of time over which the market in question extends; the influence of time being more fundamental than that of space" (Walter Isard, "The General Theory of Location and Space-Economy," *Quarterly Journal of Economics* 63:476–506).

7. Cf. C. A. Fisher, "Economic Geography in a Changing World," *Transactions and Papers, 1948* (Institute of British Geographers).

8. Stuart Daggett, *Faculty Research Lecture*, University of California, Berkeley, 12 March 1952, p. 22 (processed).

9. Cf. the work of Hartshorne, Whittlesey, Boggs, Jones, and a host of European and other geographers, as noted in Stephen B. Jones, *Boundary Making: A Handbook for Statesmen, Treaty Editors and Boundary Commissioners* (Washington, D.C.: Carnegie Endowment for International Peace, Division of International Law, 1945).

If we consider some of the subspecialties of economic geography we find that many of them are not even wholly branches of economic geography, let alone branches of economics. Consider transportation or, better still, the broader concept, implied in the French term *circulation*. This is one of my own specialties, yet I can do no better than to quote the conclusions of a leading American transportation economist, Professor Stuart Daggett of the University of California, as follows:

Most of those who have dealt with the function of transport—at least, most of those with whose work I am especially familiar—have been economists. . . . Yet the longer that I have worked with transport problems the more I have come to feel that the subject of transport, comprehensively considered, runs across the thread of human experience. History, psychology, the peculiarities of government, the associations and migrations of people as well as the fertility of land and the techniques and organization of industry are concerned with the element of space. I therefore commend the subject of transport and of space to many disciplines. I make no preferential or exclusive claim. I do feel that the permutations of space, like many other elements, may provide a starting point from which many scholars may proceed along ways which their sensitiveness and their imagination will direct.[8]

When we consider other branches of systematic human geography the situation is similar, although perhaps not quite the same as for economics or transportation. Political geography does concern itself with areas of states, capitals, and boundaries in a distinctive geographical manner. In fact political geographers probably are the experts on boundaries,[9] a topic which most political science or international relations students quite properly regard as a fringe topic of minor importance. So do political geographers, but, for geography as an areal science, we must be experts on boundaries as a part of our larger concern.

The discussion above indicates some concrete topics with which geography is more exclusively preoccupied than the other social sciences, although many topics are partly shared; the difference lies in the target center aimed at for each discipline. In other words, the contributions of human geography are probably best defined in terms of ways of thinking, approach, and methods.

THE ANALYSIS OF SPATIAL INTERACTION

I feel that the main contribution of the geographer is his concern with space and spatial interrelations. This is the common denominator of all the different types of geography whether it be the location of modern industry or such an esoteric border problem as the prehistoric place of origin and subsequent diffusion of cultivated plants. Sociology has been defined by some as the study of social interaction; by the same token geography might be defined as the study of spatial interaction. The geographer's concern with space had been compared to the historian's concern with time, although space may be more concrete. All scholars of course are concerned with space and time. What would the geographer, therefore, contribute distinctively with this point of view?

By spatial interaction I mean actual, meaningful, human relations between areas on the earth's surface, such as the reciprocal relations and flows of all kinds among industries, raw materials, markets, culture, and transportation—not static location as indicated by latitude, longitude, type of climate, etcetera, nor assumed relations based on inadequate data and a priori assumptions. I do, however, include consideration, testing, and refining of various spatial theories and concepts, some of which are noted subsequently. Furthermore, it seems to me that the spatial contribution, by definition, is particularly relevant to an area study. Following are some categories and examples of a geographer's contribution under the broad heading of space.

The Use of Maps

The geographer uses the map as his primary tool. I do not mean principally cartography, nor the surveying, drafting, or reproduction of maps, although here again is a limitation that some social scientists would logically, but mistakenly, place on the geographer's role. I mean the use, interpretation, and imaginative compilation of maps for the purpose of showing spatial interrelations. One might argue that the geographer need know no more of the drafting and reproduction of maps than the novelist need know of the mechanics of printing and setting type. The geographer also would probably take responsibility for the maps in a joint area research project and hire the draftsman, a not unwelcome service, but in itself merely a detail of the geographer's contribution.

I recall an economist once telling me that the map was a theory which geographers had accepted. The map also serves the geographer as his primary statistical tool. On it the geographer often can show correlation, dispersion, skewness, and a variety of concepts and relations better than by use of other methods. The map representation of data, therefore, often is the major part of a geographer's conclusion, not merely his tool. In any case it is an analytical device midway between data and conclusions. The map's basic contribution is to re-

". . . defining geography on the basis of what American geographers do makes it 90 percent human geography. This was one of the points I had in mind in 'Human Geography and Area Research.' This particular article did not attract much attention among geographers. . . . However, I still think that in a short space it says a lot about the philosophy of geography. It was written more to influence an intelligent outsider than geographers, but I could not help but put in some plea for reform for geographers as well. . . ." [Letter to Richard Hartshorne, December 1955]

"Thanks very much for your more than kind comments about 'Human Geography and Area Research.' Interestingly enough most of the favorable comments, or indeed comments of any kind, have come from the younger men. I have, however, received a few favorable comments from men outside the field of geography." [Letter to John W. Alexander, 9 October 1953]

duce reality to a scale which can be comprehended. Perhaps because of these virtues of the map, geography has been somewhat slow to adopt certain advanced statistical measures and mathematical formulae, which, of course, should be developed where useful.

Land Use

Geographers conventionally spend much time mapping the character and use of the land. This is important to many problems, but sometimes I feel that we do this simply because it is tangible and distinctive, in fact by now almost instinctive, much as anthropologists used to measure heads simply because it was customary. To borrow a term from science, land-use mapping is operational, but, to mix metaphors, I wonder whether the operation cures many patients? I do not, however, mean to imply that it kills the patient. After all, few patients die from having an X-ray taken or a wart removed.

If land-use mapping is done in the field by use of detailed or reconnaissance techniques, it is important to pick categories and areal units that mean something.[10] In the urban field, for example, location of furniture stores might provide a suitable category for detailed mapping in American commercial cores because furniture stores are generally distinctively clustered on the edge of the business district, whereas restaurants might be so scattered as not to warrant separate classification. Likewise some quality breakdown of residential areas is often desirable in order to determine, among other interpretations, what general land-use theory applies to the particular city, or the correlation of high grade residences with hills (quite usual in the United States, but unusual in parts of Latin America), or the relation of these areas to the plaza (a significant relationship in many parts of Latin America). In other words, land-use mapping should be geared to concepts. Here, however, the type of geographer is important. The rural geographer may contribute little in the way of concepts to urban land-use mapping and the urban geographer equally little to rural land-use mapping, although each might contribute fresh points of view.

10. Cf. forthcoming papers by Preston E. James in *Surveying and Mapping* and by Edward L. Ullman, "Advances in Mapping Economic Phenomena," *Economic Geography.*

11. In this connection I wish it were possible for geographers to incorporate some indication of intensity or quality in more land-use maps. Even a trained judgment of the field investigator classifying, for example, pasture as "good," "medium," or "poor" would be a vast improvement, even though it would expose the classification to the dangers of more subjective judgments. Subjective judgment, however, is called for even in simply indicating categories without any quality or intensity characterization, as for example distinguishing between wooded, brushy pasture and scrubby forest.

Land-use mapping is therefore a part of the geographer's contribution. For some problems it is particularly relevant, and, if done in a thoughtful and sufficiently detailed manner[11] in contrasting areas, it is a wonderful contribution to comparative general geography, far superior to mere qualitative comments or to a mechanical mapping not geared to concepts.

Flow Phenomena and Spheres of Influence

Along with determining land-use areas on various scales, in a sense a static mapping of the area, the geographer is also concerned with mapping and analyzing the flows of goods and peoples in the area—a kinetic or dynamic aspect of geography. Establishing the connections between areas is just as significant as establishing the character of the areas themselves. Origin and destination of commodities, traffic flow of goods and people, tributary areas of varying kinds around cities should all be mapped quantitatively and in varying detail depending on availability of data and the purposes of the research.

This important topic is close to the center of my own interests and is one tangible way of measuring spatial interaction, the focus of this paper. It is not considered separately in greater detail at this point, although it is a part of the remaining discussion, particularly of the following section.[12]

The Problem of Regions

The geographer should be able to contribute basically to the regional delimitation problems of the area under study, whether for the whole area in relation to others, or for parts of the area. This is merely a means to the ends of area study, but one on which much heat and little light is often generated. One of the geographer's peculiar contributions is the recognition of the various types of regions. Thus on the basis of static land-use mapping he would delimit "uniform" or homogeneous regions; on the basis of analysis of spatial connections and flows he would delimit "nodal" (or organizational or functional) regions; and, if necessary, he could work out in consultation with others a composite set of regions.

As every geographer knows, the purpose of the study controls the type of regions to be set up. Nevertheless, there are often compelling multiple-purpose regions recognizable for many purposes in their own right. An example might be the Prairie Provinces of Canada—a flat prairie, with relatively thick soil cover, growing wheat and other crops, and sustaining a rural population—in contrast to the neighboring Laurentian Upland on the east—a rocky area denuded of soil, producing trees, with mineral exploitation here and there, and with virtually no farming or rural population. Here, however, I would remind the geographer of an obvious, but perhaps, nevertheless, profound fact, namely that a region

"It is rather interesting to contrast your education and mine. You started out broadly and became a geographer. I started out as a geographer and then bumped into the cold world afterward and had to broaden. . . . My telling them [colleagues] what geography is does not make it seem significant and important when I get through, and yet I am a dedicated geographer. [Letter to Wellington D. Jones, 29 January 1953]

12. For a stimulating philosophical statement of this concept see P. R. Crowe, "On Progress in Geography," *Scottish Geographical Magazine* 54:1–19. For some applications note: R. S. Platt, "Reconnaissance in Dynamic Regional Geography: Tierra del Fuego," *Revista geografica do instituto Pan-Americano de geografica e historia*, Rio de Janeiro, vols. 5/8, nos. 13 and 24 (1949):3–22; or my own studies: *Mobile: Industrial Seaport and Trade Center* (Ph.D. dissertation, University of Chicago, 1943); "The Railway Pattern of the United States," *Geographical Review* 29:242–56; "Rivers as Regional Bonds," *Geographical Review* 41:210–25.

"You will notice that I tend to define geography fairly closely to the science of distribution definition. I would freely admit that this is not all there is to geography. However, to me science of distribution implies more than mere static location—Hettner, Hartshorne, Sauer, Schlüter, and others to the contrary notwithstanding." [Letter to Richard Hartshorne, 12 September 1952]

exists around each person and place.[13] Thus someone living on the border of the Prairies and Laurentian Upland would consider that neighboring parts of both these areas were his region.[14] In other words, an underlying characteristic of regions is the multitude of overlapping regions centered around individuals, one reflection of the very real friction of distance. (A homely example is provided by a Pole in Chicago who was asked to delimit his neighborhood. He answered that his neighborhood was as far as he was "gossiped about.")

Two corollary tendencies ensue from this fact: (1) a dense core of population tends to produce a more definite region inasmuch as the overlapping boundaries around this core tend to coincide for many individuals instead of for one or a few; and (2), conversely, in zones of sparse population or barriers to movement, a similar multiplying of boundaries tends to pile up because of the relative lack of connection across the blank spot or barrier.

In many instances the very difference between "natural" regions acts as a bond, whether it be on the local scale of an integrated valley farm with hillside pasture and valley-bottom feed crops, the giant city living in part on the interchange between unlike areas, or a large region such as the Pacific Northwest with bonds across the Cascades connecting the wet, forested ocean front with the dry east. The mere flow of goods, however, does not create a region. One of the two or three heaviest tonnage movements in the United States is the flow of coal by rail from West Virginia and contiguous Kentucky to and through Northern Ohio, yet no one would consider southern West Virginia and northern Ohio as parts of the same region for most purposes.

13. This is one type of "nodal" region in contrast to the Prairie Provinces, a "uniform" region. For definitions see forthcoming section on regional geography by D. S. Whittlesey and other members of Regional Committee, Association of American Geographers, in *American Geography: Inventory and Prospect* (Syracuse University Press, 1953).

14. One has merely to note the extra trains listed in the Canadian Pacific and Canadian National Railway timetables between Winnipeg on the Prairies and the forested lake resorts about one hundred miles east in the Laurentian Upland at Kenora and Minaki, to observe this tie.

15. Cf. the review of Baulig's *Géographie universelle* volume on North America and the references to German pleas for use of this principle in W. L. G. Joerg, "The Geography of North America: A History of its Regional Exposition," *Geographical Review* 26:654.

Basic Support and Character of Areas

In studying an area the geographer wants to pin down its basic occupations and resources. Basic employment data would be calculated from census and other sources and the underlying supports or sparks responsible for development indicated and measured, whether the forests of western Washington, the oil of west Texas, or the climate of southern Florida.

This concern with basic support is related to the principle of the "predominant characteristic" employed by French and other geographers in regional studies.[15] If I did nothing else during the war, I am glad to have insisted on having the introductory paragraph to joint military regional studies start with a summary of the basic characteristics of the area differentiating it from others, rather than using the dull and uninspired method of bounding the area by its neighbors and by coordinates, an almost universal practice in military circles.[16]

Geography and Areal Differentiation

Up to now I have not referred to geography as the study of the significance of areal differentiation, the current catholic definition of geography after Hartshorne, Hettner, James, and others. This concept is implied in the spatial or "science of distribution" point of view. I cannot accept areal differentiation as a short definition for outsiders because it implies that we are not seeking principles or generalizations or similarities, the goal of all science.[17]

The concept nevertheless has great value as a sub-concept and is the justification for the area approach. If all places were the same, a ridiculous assumption, there would be no area studies and no geography; parenthetically, if the natural environment were the same everywhere, an equally ridiculous assumption, there would still be geography, because of the advantages of specialization and economies of scale, to say nothing of the differences in culture that would create spatial patterns and contrasts. In fact it is instructive to imagine a completely homogeneous natural and cultural setting in order to formulate theories of settlement and occupational distribution, as witness the central place concept for distribution of settlements, von Thünen's *Isolierte Staat* rings of land use, or the general "gravity" interrelations of some human objects in earth space.[18]

An example of the usefulness of the differentiation concept is provided by Southern California and Florida. The whole development of these areas is largely a response to the differences between them and the rest of the United States. These are the only two significant areas of subtropical climate in the United States; this characteristic, coupled with their location in the United States, is largely responsible for their growing citrus fruit, even though slightly on the cold side of the optimum, according to Ackerman,[19] and, even more, is responsible for the astonishing increase in their population in the last two decades, as I have noted in another paper.[20] Their differences have always existed, but did not become significant until the United States reached a certain level of economic and technical development, which created footloose people and industries able to take advantage of desirable liv-

16. See my statement in "Lessons from the War-Time Experience for Improving Graduate Training for Geographic Research," Report of the Committee on Training and Standards in the Geographic Profession, *Annals of the Association of American Geographers* 36:207.

17. Cf. the similar view of Eugene Van Cleef, "Areal Differentiation and the 'Science' of Geography," *Science* 115:654–55.

18. For a discussion of Christaller's central place theory see: Edward L. Ullman, "A Theory of Location for Cities," *American Journal of Sociology* 46:853–64; for a presentation of some gravity, potential, and other models, see: J. Q. Stewart, "Empirical Mathematical Rules Concerning the Distribution and Equilibrium of Population," *Geographical Review* 37:461–86.

19. Edward A. Ackerman, "Influences of Climate on the Cultivation of Citrus Fruit," *Geographical Review* 28:289–302.

20. Edward Ullman, "Amenities and Regional Growth," forthcoming in *Proceedings of the XVIIth International Geographical Congress*, Washington, D.C., 1952. Abstract published in *Abstracts of Papers* (publication no. 6), Washington, D.C., 1952, p. 92.

". . . I have preached that we should tie in with other scientific thought, create theory, particularly spatial, think geographically, penetrate into social rather than physical aspects, pooh-pooh areal differentiation or regional synthesis as the unquestioned goal, stress connections rather than differences, etc. etc." [Letter to Edward Ackerman, 9 August 1963]

ing conditions. The significance of their difference is thus a product of space and time—location in the United States and development during the current period of American technology.

Methods

Some of the methods used by the geographer have already been indicated; only a few additional comments need be made. The geographer not only employs field mapping, but also relies heavily on interviews and local documents at the primary level. Interviews are commonly depth interviews with actual performers of the operation, and are therefore enormously useful. Random sampling questionnaires have been little used, but might also provide a useful supplemental tool. Local documents include courthouse and business records, Sanborn maps, telephone message flow, highway traffic density, and a variety of distinctive items. Sampling is also applied here, although not generally in a mathematical, statistical way nor is this often necessary. One central problem of geographers, for example, is to plot the distribution of certain items, such as charge accounts or the home addresses of automobile license plates, to indicate a sphere of influence. The rule of thumb procedure is to plot enough so that additions do not particularly change the pattern. Several hundred rather than ten thousand are generally sufficient.

Assessment of the Natural Environment

Space prevents detailed consideration of many other contributions of a geographer, such as those in the physical and resources fields, in which many geographers can make a large contribution and in which most geographers can make some. Assessment of the natural environment and its interrelation with man is one of geography's concerns, although geographers realize that the environment is essentially neutral, its role being dependent on the stage of technology, type of culture, and other characteristics of a changing society. This does not, of course, mean that man lives in a vacuum.

Assessing natural environment is the role which many social scientists have suggested they would like geography to be responsible for. This is neither so easy a task as it appears to be nor does it accord precisely with the concept of geography here presented. After all, it makes little difference whether a given industry is located where it is because of its relation to natural site qualities or to the presence of a railroad. The latter is a spatial factor just as much as the former. Furthermore, assessment of the natural environment depends on the use intended. A mountain pass is not the same thing to goats, horses, canal boats, steam engines, diesel locomotives, automobiles, airplanes, pipelines, electric wires, telephone lines, or radio; nor is fertile soil the same to the Japanese farmer and the Amazonian Indian.[21] And finally limitation of geography to this role reads it out of the social sciences and leaves untreated many topics which geography finds needing attention, unless the other social sciences themselves take over the study of these topics. Many individual workers have been doing just this in brilliant fashion, but such action is hardly likely on a large scale because the cores and approaches of the other disciplines are naturally oriented in other directions.

In spite of the difficulties enumerated above, measuring the role of the natural environment is one of geography's key contributions. In fact perhaps the two biggest problems to which the geographer is expected to contribute are directly related to this theme: (1) the proper development, conservation, and best use of resources, to which the resource geographer perhaps contributes particularly a synthesis of resource considerations; and (2) the question of national power and international relations, including the economic and military strength and strategic relations of nations and areas. This latter is a large and difficult topic beset with dangers and one on which I do not specifically dwell in this paper. Suffice it to say that geographers have made and are making contributions to both problems, with more effort devoted to the former.

GEOGRAPHY'S ROLE IN AREA STUDIES

Geographers are obviously accustomed to making area studies; consequently, it is common sense that this experience alone enables them to make a contribution. In fact one scholar, Jacques Barzun, who has little use for area studies as a discipline (a conclusion he reached on the basis of experience in a war-time area-language program at Columbia), states:

"Yet there was one valuable discovery made in area instruction that no one would now dispute about. The solidest thing about an area is its geography, and our people is singularly ignorant of any but its own. . . .

. . . It is not a subject for infants only, since everything that happens happens in space. Military strategy, industry, commerce, communication, political feeling, art, and science have locus and

21. The timing of use in relation to the stage of technology also obviously is relevant. Thus a hill was more of an obstacle to railroads fifty years ago than it is today with the use of revolutionary earth-moving equipment. The original settlers of some of the middle west sought out wood and water for obvious reasons and thus settled the stream valleys and ignored the more fertile, level prairies covering most of the interstream areas. Later comers were able to exploit the more productive prairies, once better steel plows were perfected to turn the tough prairie sod, partly because of: (1) the perfection of drilling which enabled wells to be bored deep enough to reach water not accessible to the shallower dug wells previously used, (2) the provision of efficient windmills to pump the water (now recently replaced widely by motors made possible by rural electrification or the gasoline engine), and (3) the manufacture of barbed wire which cut down the amount of wood needed for fences.

momentum. To judge and take part in them, the channels that link them through three elements must be known. And from the earliest days spent in school it would help reduce that disembodied abstract feeling about subject matter if its details could be fastened down to their point of origin." [22]

These comments are by an historian, not a geographer, but may well mirror that particular historian's early training in France where geography is a more integral part of all learning than in this country. I do not, however, quote these comments in disparagement of the contributions of others, particularly of area-minded specialists, such as human ecologists or land and location economists; quite the reverse is my feeling and I am sure the feeling of the majority of geographers.

Because of the geographer's experience in area studies as well as perhaps because of the "basic" nature of geography, the geographer's contribution may be particularly relevant to the formative stages of an area investigation. I have noticed that geographers often make a quick start on such assignments. They are accustomed to dividing areas into units and topics for investigation; their techniques enable them to start field and library research expeditiously. This is the measurement stage. For applications some of the other social scientists may be more useful than the geographer, somewhat as the engineer or the medical man applies the basic research of the pure sciences.

Area Development as a Purpose of Area Research

The purpose of an area research study has only briefly been mentioned up to now; obviously this is all-important and the contributions of the various specialists vary with the purpose. [23]

Often, in the absence of some specific purpose for an area study, the implied goal is area development of some sort. This appears to be the broad mission of the Area Division of the Department of Commerce; it evolved naturally as the center of interest of a group of regional economists meeting in California in the spring of 1952. It is the goal of Point IV programs and of much Southern regional work. In a sense the betterment of mankind is the ultimate goal of all learning, but we realize the pitfalls of premature and biased missionary efforts in this direction. Why the betterment of development of an area and its people should be the goal of area studies more than of other academic studies is somewhat of a mystery, not worth spending much time on, but nevertheless a fact.

Basic to this purpose is determination of the present stage

22. Jacques Barzun, *Teacher in America* (Boston: Little, Brown, 1945), p. 145.

23. A specialized example of an area study to which a geographer particularly could contribute was one which the Massachusetts Community Organization Service asked me, as a human geographer, to do after consulting other social scientists who neither wanted the assignment nor had the relevant techniques and concepts readily available. The problem was simply to erect some sensible sub-regions for the numerous social agencies of Massachusetts, ranging from Campfire Girls to public health units. The existing divisions were chaotic and inefficient. No one perfect set of regions could be devised, of course, but it was apparent that something could be done based on zones of influence of urban centers and characteristics of the population. However, I moved to the state of Washington and was therefore unable to pursue the investigation.

of development of the area and the explanation therefore. Advantages and disadvantages of the area in comparison to other regions and their achievements in technological and cultural development are thus key items to study. In the absence of some technological or cultural change, however, the geographer would probably reason (and prove?) that the area is as it is for good reasons. He would add a note of caution to expectations of great and sudden improvement as a result of any bootstrap-lifting operation.

Trying to find the causes for the present stage of development would run through a gamut of research investigations. For example: if the area appears "backward," is one reason for retarded development the fact that the brightest young people leave it? Is the local education poor? What effect do these factors have? The sociologist is thus faced with a problem on which he might wish to call in an educator and others for help. In most cases, however, an area study would not be final; it would do what it could and, if nothing else, would formulate questions which would in themselves be of value, since they came out of a well-rounded consideration of many facets.

Synthesis

Up to now I have largely ignored the geographer's role in synthesis, the professed goal of regional geography for many.[24] To give the geographer responsibility for synthesis is on the one hand arrogant, and on the other hand tacitly assumes that the geographer can do nothing else. Neither is true, yet synthesis is important in geography and in area studies. Regional geography, and indeed, much systematic geography, is a product of the interrelated character of phenomena on the earth. The problem is so difficult that the geographer scarcely knows where to start; everything is related to everything. Even the systematic geographer studies his specialty in relation to something else—to the areal setting. Otherwise the study has no significance.

One of the values, then, of an area study is to recognize this synthesis, or the interrelated nature of areal phenomena, the *zusammenhang*, as German geographers call it. Using a team of specialists should therefore provide the dual advantage of: (1) deeper penetration into topics and thus a more sophisticated product, and (2) a richer and fuller synthesis and approach to reality, because of a more balanced consideration of all factors.

24. I agree with Ackerman and others that the day has passed when a geographer can be a generalist, even a regional generalist, although some appear to do well with this approach, but I suspect that part of the measure of their success (other than possessing superior minds) is the fact that they are also fairly good topical or systematic specialists in a small number of fields. A systematically trained geographer in one or two fields (as for example transport, industry, mining, or various aspects of urbanism or resources) can penetrate far more deeply and ask much more meaningful questions of an area than can the untutored generalist. Even Ackerman regards the totality of geography, however, as the final goal, although we may never live to see it reached (Edward A. Ackerman, "Geographic Training, Wartime Research, and Immediate Professional Objectives," *Annals of the Association of American Geographers* 35:121–43).

Policy and Theory

As a final point mention should be made of the contributions of geography and area research toward policy and theory. Up to now I have concentrated on developing an understanding of the area, a worthy objective in itself. If policy is interjected, danger of losing objectivity is possible; in fact many area studies conducted by local and non-scientific groups have this great defect. Here the geographer can contribute objectivity by constantly comparing the virtues and defects of the region under study with those of other regions, a logical outgrowth of the spatial, or areal, differentiation outlook of the geographer. This is seldom done, mainly because we have so little in the way of studies in other areas pointed to comparison with the problems of a specific area under study. As our subject advances we should presumably have more such problem-oriented studies to use for comparison. Thus benefit-cost ratio studies of irrigation in the West could be compared with similar studies of drainage in the Southeast; the effect of size of market and operating unit—the economies of scale of all sorts—found in various size regions in the United States and elsewhere could be established and form a base from which to measure the importance of this factor where relevant in other regions under study. Obviously it will take more than geographers to provide this sort of ultimate benchmark. In the meantime some progress can be made even on the basis of our present inadequate comparative knowledge.

Equally significant are comparisons between the merits of alternative uses of government aid for development. To take an extreme example: would the nation or region obtain greater benefit from building superhighways or from doubling schoolteachers' salaries? Analysis of this sort is so difficult as almost to defy execution, yet the problem persists. Science and social science in time, I am sure, can contribute something to this basic policy problem of alternative development pointed toward the solution.[25] It is not enough to leave the solution exclusively to legislatures or philosophers.

Scarcely less difficult is the development of area or spatial theories, some of which have been hinted at previously. Here the contributions of interested representatives from a variety of disciplines should be most fruitful. Spatial concepts cannot be made in a vacuum; they must apply to something; already they have been formulated by a surprising variety of disci-

25. Cf. George A. Lundberg, *Can Science Save Us?* (New York: Longmans, Green, 1947).

plines. Further progress derived from exciting cross testing and refining should be possible with the complete pooling of these concepts. By definition the sum total of these spatial concepts should be relevant to geography, but parts of the improved product should also contribute to some of the special or general theories of the other disciplines.

I agree with the sociologists Merton and Parsons, however, that we are not ready yet for broad universals.[26] We have already been burned by one: environmental determinism. Theories should be formulated in the "middle range," intermediate between day-to-day minor working hypotheses and master theoretical schemes. Many of these "middle range" theories, however, will apply to several social sciences and may be parts of a general social science theory. Two potential virtues of area studies should be cultivated in this process: (1) comparative data from region to region should be gathered when feasible and comparative concepts tested, and (2) the limitation of each area study to one region in itself would logically seem to facilitate "middle range" theories relevant to a manageable range of variables and yet applicable to more than a unique case.

In pursuing these goals we are not particularly concerned with the contributions of, or for, one discipline, but rather with the development of all learning for the benefit of all mankind. Nor is it our intention to lay down any rigid set of procedures and topics which must be pursued. In the present embryonic state of learning, the individual inspiration of the superior scholar is apt to give more insight than any other approach. A reasonable and genuine teaming up of individuals in area research, however, and the subsequent cross-fertilization between their insights may stimulate even further the growth of individual concepts and help to codify them into a more coherent body of knowledge.

26. Robert K. Merton, *Social Theory and Social Structure* (Glencoe, Ill.: Free Press, 1949), pp. 5, 10; Talcott Parsons, "The Prospects of Sociological Theory," *American Sociological Review* 15:3–16.

CHAPTER 3   Space and/or Time: Opportunity for Substitution and Prediction

This was the last professional article Edward Ullman wrote and was in some ways a fitting summation of many of his geographic conceptions. In it he goes well beyond Kantian notions of time and space as integral elements of thought. His presentation is closer to Einstein's theory of relativity in focusing on the substitutability of time and space, which he explores in practical, down-to-earth terms.

His treatment of time and space leads to a variety of spatial explanations. Periodic market centers are comprehensively examined, and other—as yet little appreciated—effects of time on space are mentioned. Of major importance is Ullman's reexamination of central place theory within a space/time perspective. By using central-place patterning in conjunction with periodic markets, he develops an entirely new and dynamic system of central places.

There have been few ideas on which Ullman worked more persistently than this one. The first formal presentation of his remarks was made 11 February 1971, at the University of Washington, as the first Charles M. Tiebout Memorial Lecture. The final paper, as presented below, was published in 1974 in Transactions of the Institute of British Geographers *62:125–39, and reveals much about the depth and breadth of Ullman's reading and scholarship outside the traditional bounds of geography.*

1. Personal communication from Torsten Hägerstrand. A totally different but highly subjective interpretation may be provided by Matoré (1962) who notes the increasing use

To use the earth requires the organization of space and time. They provide the setting and the coordinates, and they are interrelated: one can often be substituted for, or measured in terms of, the other. Each has several definitions which provide opportunities for new and fruitful fusion.

At first glance it might appear that this analysis is concerned with historical geography. This is not the intention. The concern is rather more with exploring concepts affecting retrodiction and prediction and specifically some comparative characteristics of space and time and the opportunities for substitution between them. In a recent pronouncement in historical geography Jakle (1971) sounds in places as though he is close to the point of view of this paper, but this is largely illusory. He does, however, state: "It is doubtful . . . that geography can continue its search for spatial understanding by ignoring the integral dictates of time and space as a natural unity; thus have geographers come to focus on the processes of spatial organization through time." Prince (1971) in his analysis of historical geography also indicates that historical geographers are looking at the topic differently. Nor does the concern have much in common with Isard's (1971) economic and social treatment of space-time, which appears rather to be a sophisticated extension of conventional metrics. The fact that both were started independently does indicate that the topic is of emerging importance to explorers of human terrestrial space.

One example of space-time substitution or interdependence is provided by Hägerstrand, the Swedish geographer. He hypothesizes that each individual has a space-time envelope. A count of news items appearing a hundred years ago in the newspapers of Sweden indicated much more time-depth then than in today's newspapers, but, conversely, much less spatial extent in their coverage of news. Today, with improved communication, transport and contact, news covers more spatial extent and is less local, but also has less time-depth or historical coverage. The need to cover more space has cut down time: hence the notion that an individual has a rather fixed space-time envelope which, if expanded in one dimension, must be cut in the other.[1] Further subjective interpretations are provided by Merloo (1970) and Matoré (1962).

41

Another subjective observation is the statement by Illick (1969) referring to the American West: "Since the fundamental dimension of history is time, Americans have felt short changed and they have compensated for this deficiency by attaching importance to space." In looking at the American frontier, F. J. Turner's incorporation of space and historical thought must be seen as both merit and defect. He also wrote: "The West looks to the future, the East toward the past."

DEFINITIONS OF SPACE AND TIME

In 1920, the English philosopher Samuel Alexander wrote: "All the vital problems of philosophy depend for their solutions on the solution of the problem of what Space and Time are and more particularly how they are related to each other" (Benjamin 1966:25). Light is shed on the problem by comparing the two, using one or the other as a *surrogate*. In fact this is a primary basis for measuring or understanding anything—to compare it to something else, even including the dangerous use of analogies and metaphors. Thus Costa de Beauregard (1963) asserts: "L' équivalence entre l'espace et le temps [est le] premier principe de la science du temps,"[2] while Kummel (1966:42), an anthropologist, professes the opposite: ". . . spatial analogies represent nothing less than the death knell of thought concerning time." Thus light may be shed on the problem also by contrasting their two views.

In general terms, space is conceived of as a passive and, to this writer, a more concrete dimension than time; time is a more active and more mental construct. Space implies *being*, time implies *becoming*. Leibniz, the first replacer of Newton, said: ". . . space is the abstract of all relations of co-existence. Time is the abstract of all relations of sequence" (Feather 1959:39). Two differences between space and time therefore are: (1) time is not reversible, whereas space is; (2) since time alone deals with sequence, there is no such thing as future or past space. Under other viewpoints it will be shown that these differences need not apply.

TIME, ESPECIALLY RELATIVE

Instead of conventional measures of time, *change* appears to be the key. If there is no change, there is no meaningful time, no history. Compare this notion to the famous French historian Marc Bloch's idea of history as the science of "man in time," "the science of change" (Barraclough 1970:57). Without change, there is no time, except cyclical, repeatable, similar events. The following quotation from a simple travel book

of spatial metaphors (as in Satre's *Le mur, situations, les chemins de la liberté*) and says it is not by accident, but reflects one of the most profound sentiments of our times and quotes Pichon (*Le rêve et l'existence*), the most extreme defender of time as the most intimate concern of our existence, in contrast to space, which appears as an empirical construction of our spirit. The spatialization of thought, he says, corresponds to placing time in parentheses.

2. Costa de Beauregard prefers temporalization of space rather than spatialization of time, noting that several authors have said that the fusion of the two favors time. Needham (1966) discusses the notion of compartmentalized and continuous time in China (p. 100); Tillich's notion that space (passive) predominates over time in the Indo-Hellenic world, but that time (active) predominates over space in the Judaeo-Christian and Chinese worlds (p. 128); that causality relates to time, that the Hebrews were the first Westerners to give a value to time (p. 128); and that motion leads to space and time (p. 94).

"It is interesting to see how we are influenced by our contacts with other scholars in other disciplines. . . . It seems to me that we have to have such awareness, but unfortunately geography has sort of held aloof from the general intellectual stream." [Letter to Stephen B. Jones, 28 September 1953]

"In all the concern over the iniquities and stupidities of international and domestic air rates, I have seen little mention of the problem of charging more for a short time trip than a long time trip. For example in going to Paris for a meeting last Thanksgiving it would have cost me $450 if I went for more than two weeks, but $700 if I went for less. I was conscientious and went for one week since I had so much to do here. As a result it cost someone, presumably the taxpayers, $250 more. This is not right. Years ago we passed the 4th section of the ICC Act to make it illegal to charge more for a short haul than a long haul. We should do the same thing for time. Morally what is the difference?" [Letter to the Editor, Seattle *Post-Intelligencer*, 27 September 1971]

(Rowe 1955:226) about the town of Ondannoa, on the Bay of Biscay, illustrates the point: ". . . you live in a timeless past where the days of a man's life are measured by the diurnal tides. The town, the boats, the methods of fishing, are all *unchanging*."

Even for the concept of change to operate, it is also necessary to have some unchanging, constant cyclical features; otherwise no change would be apparent. Both order and change are necessary for time to be measured, as Schlegel (1966:501) notes. A completely unchanging world or a completely chaotic world would have no time.

In physics, as in other fields, cause and effect could also be considered to replace time, but the whole notion of time is somewhat up in the air now, since time and cause and effect may operate differently at the physical three levels of subatomic, or very small elementary particles, middle range thermodynamic, and large cosmological. In other words, a process may be time reversible, and hence the notion that time itself may be reversible; effect-cause might replace cause and effect in subatomic physics.

In physics and science or logic and mathematics, however, it has been pointed out that equations or axiomatic laws describing processes are tenseless and hence some of the so-called time reversal is simply the way laws are written. Or it can simply be nature—as Crowe stated (1965:12): "How can we be confident in pinpointing cause and effect in the circulation of a continuous medium like the atmosphere?" Before the laws start, however, an initiating factor, a contingent circumstance is required. Time is the independent variable, the contingent or initial condition. This, in a sense, is time. Hence the notion that time may be more active than space.

There are at least two other important characteristics of time; continuous, flow, or *durée*, as Bergson calls it, as contrasted with compartmentalized, boxed, or, as Bergson terms it, "spatialized" time (Costa de Beauregard 1963; Needham 1966:94, 100, 128, 130). The latter seems to imply that space is not continuous; however, "spatialized" is probably used because space appears more concrete and hence capable of being segmented or bounded more readily than time. This is perhaps doubtful; as geographers, we know that nature appears to abhor a sharp and definite boundary to settlement.

SPACE, TIME, MOVEMENT, AND PROCESS

As for space there is no particular agreement as to what the key concept, if any, is. *Movement* is suggested as the key word, although this does not fit Leibnitz' coexistence in precisely the same way that change does sequence. However, without movement, whether tangible or intangible, there are no spatial relations, the main object of study in much of geography. Mere coexistence in space without any effect of one body or process on another is like time without change, or rather like change not producing other changes. Hartshorne (1939:242, 414, 464) implies that even in regional geography the test of significance in selecting items to study is based on their relative effect on other phenomena or distributions. Likewise, since the substitution and similarity between space and time will be emphasized, it is tempting to define them similarly. Thus movement could be thought of as *change* in *space*, and change as *movement* through *time*.

One difficulty in using movement as the key word for space is that movement is also equally related to time. In fact movement is commonly regarded as an attribute of time, although Aristotle appears to relate it to both. This may be precisely why it should be used as the key word for space—to bring the flowing, kinetic notion of time into space, to make space more meaningful. For a geographer, the end result of movement is spatial relations, and only indirectly time relations. An assumption which geographers make, often very fruitfully, is to start with the thought that the closer things are in space, the closer are their relations. The relations of course are not so simple; cities, oceans, and various complementary and transferable features modify them. A similar notion occurs in the writing of some history as Haber (1971:290) notes: "Putting events in a chronological sequence is a legitimate method of descriptive historical narrative, but all too often the historian has an unstated theory of causation which leads him to select only those events as significant which fulfill the theory and then by stringing them on a chronological thread lets their sequential time order pass as a causal order."

In geography it is known that many events which happen in the same place and time may be causally unrelated, although various probability inferences may be drawn. Obtaining actual measures of movement or connections makes the inferences more certain, but by no means absolute.

Finally, space and time seem to be discovered by movement, especially by *locomotion*, as Whitrow (1966:589) notes. The individual needs personally to move around, to change location, to discover space. The initial time sense in children also develops, according to Piaget (1966:202), from comparing velocities or the space-coordinates of *changes* of *place*. For touch and vision, Lashley (1961:192) emphasizes the need for movement in order to perceive space.

Some would go so far as to conclude, as does Blaut (1961): "Relative space is inseparably fused to relative time, the two forming what is called the space-time manifold, or simply *process*. Nothing in the physical world is purely spatial or temporal; everything is process." Amos Hawley (1950:288) says, "Space and time are separable from one another only in abstraction." Long before Einstein, the relativistic revolution had begun to discard Newton, so that Poincaré at the beginning of the twentieth century, could say, "Whoever speaks of absolute space employs a word devoid of meaning" (Blaut 1961:2).

PERIODIC MARKETS AND FAIRS
AS EXAMPLES OF SPACE-TIME SUBSTITUTION

Perhaps the best geographical examples of the substitution between time and space are provided by periodic market towns, a standard settlement feature stretching from West Africa through Morocco across Africa and Asia to China, Korea, Latin America, and other parts of the underdeveloped world (fig. 3.1).

In parts of China well into the twentieth century, as well as today, small towns and villages did not operate except on market days which came at intervals of three, five, and up to nine days (Spencer 1940; Skinner 1965; Berry 1967). Between market days, the village was a virtually unpopulated shell, with nothing to eat in restaurants save rice, according to Spencer. On market days, in contrast, the village teemed with merchants who made rounds from one village to another, and peasants who came in to buy, sell, or barter their produce and goods. The peasants generally walked and only footpaths were available. The necessity of bringing the service close to the customers was compounded by the poor transport. In places where the population was dense, the markets tended to operate at shorter intervals, and be closer together. By operating only at intervals, it was possible to bring the market closer to the rural populace. About one-third, one-fifth, or one-ninth as much population would be required to support a market operating once every three, five, or nine days, as one which operated daily. In most cases, the peasants would also visit several different markets on different days, thus interjecting a competitive or comparison shopping aspect. Others have noted the same phenomenon elsewhere in the world; Dresch (1939:43) indicated that natives go to several *soukhs* in Morocco because the prices are different. One alternative to periodic markets would have

been the smaller, isolated shops or crossroads stores characteristic of some cultures and regions—outlets close to the customers, but without the variety of periodic markets. Periodic
markets brought services superior to the smallest unit of a
central-place scheme close to people by substituting long time
to allow for short-travel distances.

In Morocco, the markets are called *soukhs* and characteristically operate at weekly intervals (Mikesell 1958; Fogg 1935,
1936, 1940, 1941; Berry 1967). In fact, the *soukh* was named
after the day of the week, so that there is a repetition of
place-names all over maps of Morocco ("Soukh el Thursday,"
etc.). They are regarded, according to Mikesell, as "events"
but not settlements—points in time rather than points in
space. In contrast to China, in some parts of Morocco more
people apparently came on animals—horses, donkeys,
camels, etc. This better transport correlates with the wider
spacing and larger size of many of the markets than in parts
of China. *Soukhs*, according to Fogg, could serve up to several
thousand persons on market day, and yet consist only of
temporary stalls and tents erected in the morning and moved
out in the evening. (This appears to be the logical origin of the
phrase, "The Arabs fold their tents and silently steal away
into the night.")

Apparently in Morocco, and probably also in China, the
periodic markets are declining and are being replaced by
permanent, full-time, operating towns. This should in part be
a response to improved transport—it takes less time to go
longer distances and hence the advantages of continuous
markets can be exercised. Increased volume of buying and
commercialization, as well as the desire of administrators to
have a more fixed population in order to carry out duties such
as tax collection more efficiently, are apparently also features
of the decline of periodic markets.

In Korea, Stine (1972; also Hay 1971) has noted that mobile
firms—pedlars or truckers with supplies—set up periodic
markets on regular schedules. He notes: "Transportation
costs are generally high in Korea. Since the majority of Korean consumers travel to central places on foot, the transportation cost involves expenditure of human energy. At the
same time, opportunity costs are apt to be high. Foot travel is
slow and the time spent in going to market means time lost
on the farm." He concludes pithily: "The consumer, by sub-
mitting to the discipline of time, is able to free himself from
the discipline of space."

Currently, extensive work on periodic markets is being
done in Africa, following on the earlier work printed in

TIME OF MARKET AND SIZE OF TRADE AREA
Increase in Market Frequency = Increase in Trade Area Required
(Homogeneous Plain, Hexagonal Trade Areas, Market Size Constant)

A. TRADE AREA REQUIRED FOR MARKET MEETING

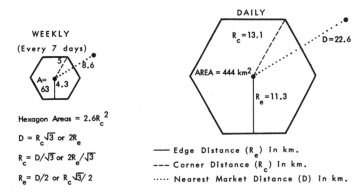

Hexagon Areas = $2.6R_c^2$

$D = R_c\sqrt{3}$ or $2R_e$

$R_c = D/\sqrt{3}$ or $2R_e/\sqrt{3}$

$R_e = D/2$ or $R_c\sqrt{3}/2$

——— Edge Distance (R_e) in km.

- - - Corner Distance (R_c) in km.

····· Nearest Market Distance (D) in km.

B. TRADE AREA AND SPACING NEEDED TO INCREASE FREQUENCY OF WEEKLY MARKET

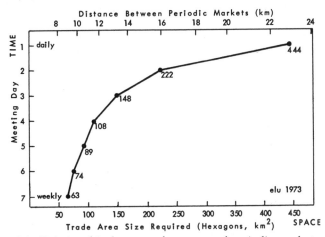

Fig. 3.1. Relationship between frequency of periodic market, spacing, and size of trade area

Bohannon and Dalton (1962). Work by Smith (1970, 1971) and by Hay (1971) are representative of this effort, and Bromley (1971) has an excellent review of many other studies. They all confirm the observations noted above and present fuller quantitative measurements, indicating, for example, the compatibility of the separation of schedules of periodic markets in time and space with central-place theory. Smith (1972) notes, for example, that in some twenty-five regions all over the world, the distances apart of periodic markets meeting on the *same* day is about twice as great as those meeting on *different* days.

IDEALIZED SPACING OF MARKETS IN TIME AND SPACE

The theoretical relationships between temporal and areal spacing of periodic markets, based on idealized hexagonal trade areas, are summed up in figure 3.1 (a and b). (In reality the maximum distances would be somewhat greater, since idealized hexagons are seldom found.) Thus if one assumes that a 63 km² trade area is required to support a weekly market, seven times as much area would be needed to support a daily one. This would mean an increase in travel distance, from the edge of the trade area to the market, from 4.3 for the weekly market to 11.3 for the daily one; from the apex of the trade area the increase would be from 5 to 13; and the increase in the distance between markets would be from 8.6 to 22.6 km. If the market were closed one day a week only six times as much trade area would be required—398 km².

The assumption of five kilometers to the corner of the weekly or smaller market area corresponds closely to Christaller's assumption of four to five kilometers to the corner of his smallest hexagon, approximately the distance one can walk in one hour and empirically the normal service area limit in southern Germany in 1930 (Christaller 1933; Ullman 1941; Skinner 1964–65).

Perhaps the most significant point is that if the distance to the corner in the weekly market is 5 kilometers, but 13.1 to the daily one, this increase would require almost three hours walking, presumably more than a tolerable limit; but the journey would take only about half an hour in a bus driven at 26 kilometers per hour on a rather poor road. If one assumed half an hour of waiting and walking time for the bus, still only about an hour would be required. Thus, with roads and motor vehicles, a daily market could operate with the same maximum travel times as a weekly market before roads. Examples of other possible relationships are as follows:

1. If population density doubles, seven-day markets could double in size or operate approximately every four or three days.
2. If purchasing power doubles, seven-day markets could double in size or operate approximately every four or three days.
3. If purchasing power doubles and population doubles, seven-day markets could approximately quadruple in size, or operate approximately every two days.
4. If transferability (transport cost and time) is reduced by

one-half, the trade area of a seven-day market could quadruple in size (from 63 km² to about 260 km²—fig. 3.1), or seven-day markets could operate at least every two days at the same size, or four-day markets could operate daily.

OTHER SYSTEMS: PEAKS, WORK STAGGERING, ETC.

At other scales, further systems of organizing space and time operate or have been proposed. In Canada, for example, some small branch banks in areas of poor transport and light population are open only at weekly intervals or less, although these periodic types are disappearing as transport and agricultural technology improve and agricultural employment declines. The same is true of many other facilities and services elsewhere in the world. Particularly appealing are proposals to stagger opening and closing hours of work in order to spread out peak traffic. This is experimented with in some cities, but not too widely in the world.

Dahl (1959), the Swedish geographer, noted, particularly in studies in Sweden, that generally the larger the city, the shorter the lunch hour, because in a big city the distances were too great for workers to go home to lunch. Alternatively, the more concentrated the city and the better the transport, the more chance there is of a long lunch hour in some cultures such as those of Madrid or Barcelona in Spain. Still another system was provided by the abandoned experiments of the Soviets in the 1920s and 1930s for a continuous work week with every fifth day off on a staggered basis (Moore 1963:122).

To avoid the peaks entirely and operate many facilities more continuously, however, would destroy the very reason for coming together to trade and interact, or would interfere with family life on weekends, although further experimentation could be tried. As Karl Deutsch notes (1961), synchronized hours of work permit a larger range of choices than staggered hours. Timing options are available, but within limits. All of these are related to daylight and darkness, eating and sleeping, cultural and biological conventions, and particularly changes in technology.

Finally, of considerable significance are the observations of Stine (1972) who points to the fact that other systems of more or less primitive land uses also are founded on space-time substitutions—nomadic herding, primitive shifting cropping (Milpas, etc.) and perhaps livestock ranching and transhumance. Particularly apposite is the notion of resource drawdown or renewable resource, as fertility is periodically exhausted after cropping and renewed after fallow in shifting agriculture. Conceptually this is similar to the draw-down and regeneration of purchasing power in the tributary area of a periodic market after opening and closing. Space-time substitution is widespread over the earth's surface, and not simply an oddity.

COST OF MOVING IN TIME AND SPACE

In assessing the mechanisms of periodic markets and other spatial-temporal systems, it is desirable to compare, if possible, the passage of time with transferability over space. What is the effect of the relative costs of traversing each? If one thinks in terms of production, as Böventer (1962) says: "The role of space in production is comparable in many ways to that of time." By this he means, presumably, that high transport costs and high interest rates both foster less favorable production methods. High transport costs restrict the size of the market and consequently the scale of production and often of the amount of plant, machinery, and overheads that can be afforded. Similarly, high interest rates discourage capital investments and large-scale, modern equipment expenditures or replacements.

Both transport costs and interest rates are characteristically high in underdeveloped countries and both work to discourage development. Transport, especially road building, and subsidized interest rates or guaranteed loans for capital improvements and machinery are two of the principal development programs in underdeveloped countries. Still another need in many underdeveloped countries (and in inner portions of many large American cities) is better internal security. Poor security produces risks which affect both travel, especially at night, and the interest rate on loans. Periodic markets with mobile vendors enable producers to fully utilize their small capital, and customers to make short trips which precisely fit into such conditions. When transport improves and/or capital becomes available, often in the form of government installations, periodic markets are replaced by permanent ones, especially when population and income increase.

The other way to view the problem is from the standpoint of the customer; transport is expensive—the buyer generally must walk because there are no roads or he cannot afford a vehicle—and this takes time and effort, especially if the distance is great. By postponing his shopping trip until a market comes closer to him he saves much travel effort. The utility of place in this case might be said to be of more value than the utility of time. A simple comparison of costs perhaps indicates this. Freight rates as a proportion of total costs can be quite high; 40 or 50 percent is not uncommon for characteristic shipments of bulk commodities in the United States, or for even more valuable commodities moving shorter distances by primitive transport in underdeveloped countries. The exact calculations are difficult, because of so many variables, but even so (and even for marginal analysis) the transport costs appear to be much greater than interest rates which range from 5 to 15 percent per year, although local moneylenders on short terms would be apt to charge much more. However, a postponement of purchase of even a week would entail a negligible interest payment even if socially inflated. In addition, low wage rates in primitive economies may well make the value of time less than the effort of difficult movement.

Changes in technology will reduce the cost of movement, but there is no reason to suppose any consistent change in the utility of time. If anything, utility of time may go up as utility of space goes down. This presumably spells the doom of periodic markets on economic grounds.

However, three important special characteristics of periodic markets should be re-emphasized to explain their nature and persistence. First, periodic markets generally are rather large—a hundred or more sellers and stalls—and thus provide fairly extensive comparison shopping. They are the equivalent of a higher-order central place, not a low- or smallest-order market hamlet (*Marktort* in Christaller's scheme) or an isolated, monopoly outlet.

Secondly, as Bromley (1971) and others note, periodic markets appear to be patronized particularly by the poor—natives or peasants—and hence the less mobile; moreover, purchasing power per area is less. Thirdly, and of great importance, most periodic markets are heavily dominated by exchange or sale of local products. Local goods are brought and sold locally (Skinner 1964–65). Thus both supplier and customer are local, reinforcing the need for close spacing. Good (1970), for example, observes in East Africa that full-time traders come from a distance of up to fifty miles, but most part-time (local) traders selling agricultural and other local products go an average distance of only .01–5 miles.

Let us leave this complex question of comparing the cost of moving through time with space, and move on to some other tantalizing concepts, primarily of relative space and time.

DIRECTION IN TIME AND SPACE

Is time reversible? Space is, but, strictly speaking, time is
not. Sometimes, however, time is reversible when viewed in
certain ways. If change is the measure of time, then time can
be somewhat reversible in human affairs. A common example
is the prediction of some future level or event and the subsequent adjustment to that prediction—to build "effects" to
meet this future "cause"—as in the construction of a highway
or other facility which initially produces excess capacity—the
familiar "lumpiness" characteristic of public works and other
investments. Traffic will flow to the new freeway leaving
other arterials relatively empty; a few years later they may fill
up again. It is hardly correct to say that as soon as freeways
are built they run to capacity. They run to a future predicted
capacity, a time reversal. Predicting the future can also have a
similar effect to the extent that people plan ahead of time to
meet a future goal or effect (Dickson 1972).

Lags and leads in construction are characteristic of our development because of poor planning or other circumstances.
Thus most of the railroads in the United States, especially in
the less well-developed sections, went bankrupt early in their
careers. They were built ahead of their time, although perhaps not so much so in some cases if their external benefits,
such as sale of land for town sites by subsidiary companies or
windfall profits to outsiders, were considered. Even these
profits were speculative, time discounted on the future. The
Union Pacific, the first transcontinental railway in the United
States was finished in 1869, a full thirteen years before the
next one. Fogel (1960) in an exhaustive study calls it a case of
premature enterprise. It brought the future to a region before
it could support it, but distant regions paid for it for strategic
reasons, a fundamental characteristic of long-distance land
transport.

Moving to another place can also be equivalent to turning
the clock back (or forward). Benjamin Franklin observed: "A
thousand leagues is as a thousand years." By this he probably
meant that going a thousand miles into the interior of a primitive continent was equivalent to going back in time a
thousand years or more (or vice versa if one starts from a
primitive environment). The world, thus, is full of old time,
generally in distant, remote places untouched by civilization.
This is a favorite theme in literature. It is also the motivation
for much anthropological work such as in the remote interior
of New Guinea before it becomes too accessible. Santos (1972)

notes that it is the fundamental basis not only for differentiating countries and regions, but even more for the enormous problems of the underdeveloped or Third World realms.

Some generations, it has been observed, also seek to "walk backward into the future"—to consciously recreate old customs, styles, modes. Age is also a fundamental basis for the tourist industry, so much so that we have run out of supply and now build new old scenes—steam railroads, frontier towns, pioneer villages, Williamsburgs, Indian villages. Prince (1971:36) notes [that] "copying and restoration produced a counterfeit heritage which reminds Europeans of their past." The authentic past is becoming a real regional resource.

FUTURE OR PAST SPACE

What is future or past space? Under certain operations we can create its equivalent. Consider the economic base of cities where the crucial question is to measure the exports. How does one classify construction, much of which is generally local or internal, if a given city has a high employment, well above the minimum or even the average (location quotient)? In this case, would it not be sensible to think of the excess as export into future time, just as excess in other activities is export into space beyond the city? Here contributions to capital account can be thought of as equivalent to exports, certainly not a consumption *here* and *now*. If the city is rapidly growing, one can also think of some of the excess as catching up to the past requirement, and another part, more speculative, and so characteristic of the building trades, as export into the ever-anticipated future growth, especially in places where growth trends and optimism are endemic, as in southern California.

An example at a national scale is provided by Canada. From about 1900 to 1913, at a time of heavy railroad construction, it was noted that "exports were relatively reduced because so much Canadian effort was being given to capital formation" (McDougall 1968:93). Exports can thus be in space or time—through trade or investments.

SPACE-TIME TREATMENT IN LANGUAGE AND LITERATURE

Space-time treatments are widely imbedded in language and literature; this argues that they are pervasive components of our culture, worth studying and not simply abstractions of physicists or philosophers. Space generally appears to be a more tangible, less elusive concept than time, and for this reason is often used as the measure of time. Time, however, appears to be more meaningful to man himself as a felt experience. It also is used as a measure of space, although less frequently than space as a measure of time.

The greater use of spatial figures of speech thus appears related to the physical measure of time. As Ward (1961:5) says: "The passage of time is marked by change and motion and all methods of measuring time are based on some form of regular motion, by means of which the passage of time is translated into the traversing of space."

Nevertheless, time units are often used, as when one is told that a certain place is only ten minutes away. This is more meaningful to man, as are most time measures, than using a distance measure, as Post (1963) notes. Various leagues were simply measures of time needed to traverse a certain distance. With changes of technology, these time measures became obsolete, but were continued in use without any time connotation. Likewise, some time measures are simply old spatial measures, now with no spatial connotations. Nilsson (1920:281; also Sorokin 1943:191) observes that varying lengths of time periods, of old weeks of three, eight, ten days, etc., were originally associated with periodicity of markets, and we have already noted that the periodicity of markets was a human response to space.

Some examples of space surrogates in our idiomatic and literary speech today are: "he covered a lot of ground"; "he was at a certain place (or point) in time"; "his field is the nineteenth century." Writers also substitute space for time to add interest and emphasis as Faulkner in a letter to Cowley: "Life is a phenomenon—the same frantic steeplechase toward nothing everywhere and man stinks the same stink no matter where in time." The same use is made by Waugh (1961:175): "For four years they had travelled in one direction, I in another. There was a difference of eight years between us."

Examples of space-time equivalents are shown in the table at right. Not everyone would agree that they are good equivalents. Gale (1969), for example, states: ". . . it seems to me that the disanalogies between 'here' and 'now' are profound: space and time are radically different." In taxonomic terms he might thus be termed a "splitter" as opposed to a "lumper."

In any event, the space terms can be used for time more than time for space. Thus it is more possible to say here in time than now in space, or point of infinity in time, than

A DICTIONARY OF SPACE-TIME SURROGATES: A TRANSLATION OF EQUIVALENTS FOR A SPACE-TIME LANGUAGE

Space	Time
Here	Now
There	Then
Where	When
Everywhere	Always
Nowhere	Never
Farther	Further
Point	Instant
Region	Period
Origin	Invention

instant or eternity in space. Many other examples of substitution and equivalency appear to revolve around the notion that space is more tangible as a measure and hence is used more as a measure of time, but time as a concept is more abstract but more used and more meaningful to man.

Various languages and grammars have structures and words relating to space and time. An illustration is provided by Professor Farhat J. Ziadeh, Chairman of the Department of Near Eastern Languages and Literature at the University of Washington:

According to Wright (1964, p. 124), there is the concept of the *nomina loci et temporis* or nouns of place and time. The same noun pattern doubles for both the time of action and the place of action. Arabic grammarians have also called this concept *nomina vasis* because time and place are, as it were, the vessels in which the act or state is contained. For example, the word *maghrib* means both the place where the sun sets, and the time when the sun sets. Thus, it means "west" and it also means "evening" and, by extension, to the Arabs of the East it means also Morocco because that is the place where the sun sets. Likewise, the word *mashriq* means both the time of the shining of the sun in the morning and "east" meaning the place of rising of the sun. This concept is also found to a certain extent in Hebrew.

Examples of time surrogates also occur in other languages: "mezzogiorno" (midday) in Italian and "midi" in French are used for the South; "Abendland" (evening) for West or "Morgenland" (morning) for East in German; "oriens" (rising) for East and "occidens" (setting) for West in Latin.

The explicit recognition of the oneness of time and space has also been noted by novelists and poets. Andrew Marvell used the similarity with devastating effect in his famous poem, "To His Coy Mistress":

> Had we but world enough, and time,
> This coyness, Lady, were no crime.
> . . . .
> But at my back I always hear
> Time's wingèd chariot hurrying near,
> And yonder all before us lie
> Deserts of vast eternity.
> . . . .
> The grave's a fine and private place,
> But none, I think, do there embrace.

As a final example a literary friend compares the space flights and walking on the moon to man's yearning for immortality, for transcending human limits. Mastering something outside the physical environment is the closest one can come to mastering the major personal feature of time-death. Symbolically man has reached in space beyond the fact of death, but at the cost of billions of dollars, which many argue should be spent here, and not there.

THE DEPICTION OF SPACE AND TIME

In order to perceive space or time, it needs to be condensed down to scales where a large quantity of it is visible. To do this geographers look at the world in a sense through the wrong end of a telescope, not a microscope. The map is used as principal tool, both in preliminary analysis, and in appreciation of space and spatial relations, and later in the construction of special maps to illustrate findings. To condense space, of course, involves dropping detail as scale is decreased. By doing so, some detail is lost, but relations are perceived which are otherwise invisible.

What is the equivalent tool for time? Only three tools will be briefly considered: (1) games, (2) the theatre, and (3) scenarios. The problem is similar—one cannot perceive a long time span if it is cluttered with detail. For games, note the following quotation from Feldt (1972:4): "Finally the most important property of all games is their ability to collapse time and space for the players involved. Every game is constructed at some level of abstraction. Many real world decisions are reduced to background noise to be handled as random factors or ignored by the players."

For the theatre only one specialized example will be noted. The contrast between absolute and relative space-time is well illustrated by Beckett's play, *Waiting for Godot*. This is summed up in a review by Barnes (1971):

The point of Godot is that it is a play concerned with time rather than action. . . . Two tramps "have been waiting at the same place, but possibly not, for years and years. . . ." They are getting older. They smell. . . . "Nothing happens, nobody comes, nobody goes, it's awful. . . ." Beckett regards existence as that passage of time between the cradle and the grave, and man has no function but to "pass the time". . . . Most plays have plots as formal as a Senate intrigue or a game of consequences . . . that is a legitimate way to write a play, and, of course, the most common. But Beckett has chosen not to write about action, not even symbolically. His play is about waiting.

And finally the reviewer notes that the play "is infinitely helped by the stark placelessness of . . . [the] setting. . . ." The play is a non-event in absolute (empty) time and space.[3]

Scenarios, so popular in some quarters now, are another form of sequential condensation, in this case of the future which Herman Kahn (1967:263) says "forces one to deal with dynamics that the abstract misses."

3. Drama's preoccupation with time is indicated by Ungava (1971) who comments: "There is a time emphasis in every definition of drama because drama has a preoccupation with time. To draw just a short parallel with physics, time seems to be as much an attribute of drama in the literary field as it is of motion in the natural sciences. Drama is an imitation of action and action is a form of motion. It is no accident that Aristotle . . . so closely defined time as an attribute of motion. . . ." He goes on to note that poetry=time, painting=space, and dance combines the two.

CONCLUSION

This discussion of space and time can be summarized under the two headings of similarities and differences, but these similarities and differences are far more significant than mere labels.

1. Similarities: The manifold possibilities of substituting space and time in an interdependent fashion advance the understanding of both primitive and modern environments. The substitution is pervasive and fundamental, not simply an oddity.

2. Differences: Time is generally the more active and mental construct and space the more passive and concrete dimension. To make space more meaningful, some time characteristics need to be applied to it: (1) as in thinking of *movement*, both a space and time phenomenon, as the key to a more significant space, and (2) in thinking of future and past space, or of old and new time spaces on the earth, as in the contrast between developed and underdeveloped regions. For time the notion of *change* is a key characteristic and is fundamental to this newer way of comparing space and time.

Space and time can thus be stretched into each other, or substituted one for the other; viewed in one way, they are the same, while in another, they do not exist at all. Space is partly subsumed under *movement*, and time under *change*. Empty or static space or time has no meaning. Fifteen years ago I made the revolutionary, but now obsolete, observation that movement provided a "motor" for what is called in geography "situation concepts" (Ullman 1956). By the same token, change is the *definer* of time. Without change, there is no time, no history. Without movement, there are no spatial relations, the real object of study in the analysis of space.

These features may take away some of the significance of the abstract or geometrical analysis of space. It sounds almost as though one were arguing for Ritter's famous geographical definition of space first formulated in 1833: "Die irdisch erfüllten Raüme der Erdoberfläche" (Hartshorne 1939:142). This is essentially untranslatable, even in German (as is normally the case), but literally means: "the earth filled spaces of the earth's surface." More recently "dinglich" has been commonly substituted for "irdisch". Empty space, therefore, as Ritter may have implied, is almost an empty phrase. Space takes on meaning when it contains something; this is particularly true when the "thing-filled spaces" exhibit *movement* or *change*, which is the modification that one might make of

Ritter's famous remark of over a hundred years ago. Einstein's reported explanation of relativity says in a few words the same (Frank 1957:123; Schlegel 1961:149): "If you don't take my words too seriously, I should say this: if we assume that all matter would disappear from the world, then, before relativity, one believed that space and time would continue existing in an empty world. But according to the theory of relativity, if matter and its motion disappeared, there would no longer be any space or time."

Modern relativity theory as developed in physics, might be extended beyond the conventional, operational and definitional rigor of physicists to come up with a proposition that the fixed or constant item, is really the unfixed—speed, and as measured in physics, the speed of light specifically. If speed is the point from which the world is viewed, space-time is simply a continuum moving past at about 186,300 miles per second, the speed of light measured in conventional terms. Other speeds, or rather inertial velocities, are less than the constant one of light, which not only is the fastest, but always has the same velocity in any inertial frame of reference.

Since the speed of light on earth is virtually instantaneous it is at the other end of the scale from the static state of no movement at all. At the speed of light time stops. Television has the practical effect of cancelling Robb's famous and theoretically correct dictum: "An instant cannot be in two places at once." Even before television, Malin (1944:108; Wolfe 1961–62:57) wrote: "Power machines have produced so remarkable an acceleration in the rate of change that a totally new relationship has been introduced among the three historical factors of time, space and movement. Time has been a saving factor in history heretofore, not only to the single generation of individuals, but to the overall adjustment of man to change." (Compare this notion to Aristide Briand, pre-war Prime Minister of France, who, according to Chamberlain [1935:186] consciously took "time for an ally and counted on its aid in overcoming difficulties"—a sort of positive view of risk and uncertainty!)

Virtually instantaneous movement, and potentially more rapid diffusion are features of electronic communication, as contrasted with physical transportation, which is capable selectively of abolishing space, except for human contact, as a limiting factor in the size of community. Even at speeds of 1,000 miles per hour, which are now almost commercially practicable, crossing the United States would take less than three hours, the Atlantic not much more, and most present intercity trips in the world would take less than one hour. We are approaching, even in passenger transport, a speed where the critical factors are simply initiation, loading and unloading, and terminal delay and costs. Even costs are starting to be negligible, particularly in communication, but also in transportation, especially at slow speed for bulky, low-value commodities shipped by large-diameter pipelines, or by huge bulk carriers, whose capacity may be so great as to represent in only a few shiploads the total annual output of a large mine—conceptually close to the equivalent of relocating the whole mine.

The real problem of the future, then, may not be transportation or communication, nor processing or routine planning and paper work, all of which are already highly mechanized. It is perhaps simply the problem of communication overload. The terminals, in the form of human beings, are already overloaded and require all sorts of protective filters (which unfortunately do not always screen out the least desirable features—unpleasant and excessive advertising for example). Time may be more intractable than space.

All of this, together with the increasing use of people as information processors, and as service providers, in contrast with declining numbers involved in physical activity (agriculture and industry) is the basis for Webber's (1963) speculations on a future spaceless world where people can live wherever they want. But this is getting ahead of the time or story.

A final characteristic of time, beyond serving as a substitute for space, has a profound significance for geography. Time can be thought of as the independent variable, space as the dependent variable. In other words, time is the cause-and-effect element. This is the basis for both prediction and retrodiction, a time reversal. Choosing a goal means choosing an initial condition—picking an effect in advance, if one will—and then proceeding to make it work. As Watanabe says (1966:561), "The separation of action into goal and means [is] identical with the separation of development into cause and effect. . . ."

Time provides for action, for a story, for a plot. Who has not had the guilty experience in reading a novel of skipping over descriptive-geographic passages in order to follow the plot? Often it might be argued, with reason, that the skipping was done because the geography was poor. Economist friends, however, have complained that what they miss, in a geographer's otherwise good work, is a lack of action, or even more of normative ends. Explicit recognition of time can help geographers to move in this interesting direction, sensibly and objectively. Geography already is moving toward it through its heightened concern for understanding and prediction (Berry 1970:21; Abler, Adams, and Gould 1971).

REFERENCES

Abler, R.; J. S. Adams; and P. Gould. *Spatial Organization: The Geographer's View of the World*. Englewood Cliffs, N.J.: Prentice-Hall, 1971.

Alexander, S. *Space, Time and Deity*. New York and London: Macmillan, 1920.

Barnes, C. *New York Times*, 5 February 1971.

Barraclough, G. *New York Review of Books*, 4 June 1970.

Benjamin, A. C. "Ideas of Time in the History of Philosophy," in *The Voices of Time*, ed. J. T. Fraser, New York, 1966.

Berry, B. J. L. *Geography of Market Centers and Retail Distribution*. Englewood Cliffs, N.J.: Prentice-Hall, 1967.

——. "The Geography of the United States in the Year 2000," *Transactions of the Institute of British Geographers* 51:21–53.

Blaut, J. M. "Space and Process," *Professional Geographer* 13(4):1–7.

Bohannan, P., and G. Dalton, eds. *Markets in Africa*. Evanston, Ill.: Northwestern University Press, 1962.

Böventer, E. von. *Theorie des räumlichen Gleichgewichts*. Tübingen, 1962.

Bromley, R. J. "Markets in the Developing Countries: A Review," *Geography* 56:124–33.

Chamberlain, A. *Down the Years*. London: Cassell, 1935.

Christaller, W. *Die zentralen Orte in Süddeutschland*. Jena, 1933. English translation by C. Baskin, *The Central Places of Southern Germany* (Englewood Cliffs, N.J.: Prentice-Hall, 1966).

Costa de Beauregard, O. *La théorie physique et la notion de temps*. Paris, 1963.

Crowe, P. R. "The Geographer and the Atmosphere," *Transactions of the Institute of British Geographers* 36:1–19.

Dahl, S. "Duration of Lunch Interval: An Aspect of Urban Social Geography," *Acta Sociologica* 4(2):16–19.

Dickson, P. *Think Tanks*. New York: Atheneum, 1971.

Dresch, J. "Les genres de vie de montagne dans le Massif du Toubkal," *Revue géographie maroc.*, no. 1, 1939.

Deutsch, K. W. "On Social Communication and the Metropolis," *Daedalus* 90(1):99–100.

Feather, N. *An Introduction to the Physics of Mass, Length, and Time*. Edinburgh: Edinburgh University Press, 1959.

Feldt, A. G. *Clug: Community Land Use Game* (players' manual). New York, 1972.

Fogel, R. *The Union Pacific Railroad: A Case in Premature Enterprise*. Baltimore: Johns Hopkins University Press, 1960.

Fogg, W. "Village and Suqs in the High Atlas Mountains of Morocco," *Scottish Geographical Magazine* 51:144–51.

———. "The Economic Revolution in the Countryside of French Morocco," *Journal of the Royal African Society* 35:123–29.

———. "Villages, Tribal Markets and Towns: Some Considerations Concerning Urban Development in the Spanish and International Zones of Morocco," *Sociological Review* 32:85–107.

———. "Changes in the Lay-out Characteristics and Functions of a Moroccan Tribal Market Consequent on European Control," *Man* 41:104–8.

Frank, P. *Philosophy of Science*. Englewood Cliffs, N.J.: Prentice-Hall, 1957.

Fraser, J. T., ed. *The Voices of Time*. New York: Braziller, 1966.

Gale, R. M. "Here and Now," *The Monist* 53(3).

Good, C. M. "Rural Markets in East Africa," Department of Geography Research Paper no. 128, University of Chicago.

Haber, F. C. "The Darwinian Revolution in the Concept of Time," *Studium gen.* 24.

Hartshorne, R. *The Nature of Geography: A Critical Survey of Current Thought in the Light of the Past*. Lancaster, Pa.: Association of American Geographers, 1939.

Hawley, A. *Human Ecology*. New York: Ronald Press, 1950.

Hay, A. "Notes on the Economic Basis for Periodic Marketing in Developing Countries," *Geographical Analysis* 3:393–401.

Ilick, J. E. "Looking Westward," *American West* 6:50.

Isard, W. "Relativity Theories and Time-Space Models," Papers of the Regional Science Association, no. 26.

Jakle, J. A. "Time, Space and the Geographic Past: A Prospectus," *American Historical Review* 76.

Kahn, H. *The Year 2000*. New York: Macmillan, 1967.

Kummel, F. "Time as Succession and the Problem of Duration," in *The Voices of Time*, ed. J. T. Fraser, New York, 1966.

Lashley, K. S. "The Problem of Serial Order in Behavior," in *Psycholinguistics*, ed., Sol Saporta, New York, 1961.

Malin, J. C. "Space and History: Reflections on the Closed Space Doctrines of Turner and Mackinder and the Challenge of Those Ideas by the Air Age," *Agricultural Historian* 18 (1944).

Matore, G. *L'espace humain*. Paris: La Colombe, 1962.

McDougall, J. L. *Canadian Pacific: A Brief History*. Montreal: McGill University Press, 1968.

Merloo, J. A. M. *Along the Fourth Dimension.* New York, 1970.

Mikesell, M. "Tribal Markets in Morocco," *Geographical Review* 48:494–511.

Moore, E. *Man, Time and Society.* New York, 1963.

Needham, J. "Time and Knowledge in China and the West," in *The Voices of Time,* ed. J. T. Fraser, New York, 1966.

Nilsson, M. P. *Primitive Time Reckoning.* Lund: C. W. K. Gleerup, 1920.

Post, L. C. "Measuring the Way," *Landscape* 3(3):23–25.

Prince, H. C. "Real, Imagined and Abstract Worlds of the Past," *Progressive Geographer* 3:1–86.

Rowe, V. *The Basque Country.* London, 1955.

Santos, M. "Dimension temporelle et systemes spatiaux dans les pays du Tiers Monde," *Revue Tiers Monde* 12(50):247–68.

Schlegel, R. *Time and the Physical World.* East Lansing, Mich.: Michigan State University Press, 1961.

———. "Time and Thermodynamics," in *The Voices of Time,* ed. J. T. Fraser, New York, 1966.

Skinner, C. W. "Marketing and Social Structure in Rural China," *Journal of Asian Studies* 24(1):3–43; (2):195–228; (3):363–99.

Smith, R. H. T. "A Note on Periodic Markets in West Africa," *African Urban Notes,* pp. 29–37.

———. "West African Market Places: Temporal Periodicity and Locational Spacing," in *The Development of Indigenous Trade and Markets in West Africa,* ed. C. Meillassoux, pp. 319–46. London, 1971.

———. "Spatial-Temporal Features of Periodic Markets," paper presented to Association of American Geographers, Kansas City, 23–26 April 1972.

Sorokin, P. A. *Sociocultural Causality, Space, Time: A Study of Referential Principles of Sociology and Social Science.* Durham, N.C.: Duke University Press, 1943.

Spencer, J. "The Szechwan Village Fair," *Economic Geography* 16:48–58.

Stine, J. H. "Temporal Aspects of Tertiary Elements in Korea," in *Urban Systems and Economic Development,* ed. F. R. Pitts, pp. 68–88. Eugene, Oregon, 1972.

Ullman, E. L. "A Theory of Location for Cities," *American Journal of Sociology* 46:853–64.

———. "The Role of Transportation and the Bases for Interaction," in *Man's Role in Changing the Face of the Earth,* ed. William Thomas, pp. 862–80. Chicago: University of Chicago Press, 1956.

Ungava, T. "Time and the Modern Self: A Change in Dramatic Form," *Studium gen.* 24 (1971).

Ward, F. A. B. *Time Measurement.* London, 1961.

Watanabe, S. "Time and the Probabilistic World," in *The Voices of Time*, ed. J. T. Fraser, New York, 1966.

Waugh, A. *My Place in the Bazaar.* New York, 1961.

Webber, M. "Order in Diversity: Community without Propinquity," in *Cities and Space*, ed. L. Wingo, pp. 22–54. Baltimore: Johns Hopkins University Press, 1963.

Whitrow, G. J. "Time and the Universe," in *The Voices of Time*, ed. J. T. Fraser, New York, 1966.

Wolfe, R. "A Geographer Looks at Herman Kahn," *International Journal* 17 (1961–62).

Wright, W. *A Grammar of the Arabic Language, I.* Cambridge: Cambridge University Press, 1964.

PART II | Regionalization

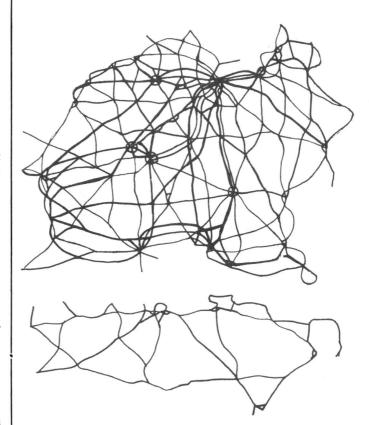

"I note that you say the Regional Concept is supposed to set the stage for the rest of American Geography, etc. Never at any time, was this the intention. In fact, I think I made it quite explicit . . . that the 'regional concept' should not set the stage—not that I am against regions. But to define geography as regional geography is asinine." [Letter to Franklin R. Stearns, 2 March 1955]

"My personal interest in geography is in terms of location, movement, interaction— situation rather than site—what one can read on a map rather than the vertical relationships between soil and agriculture, for example. . . . I would not throw out regional geography; I am still interested in it, but there is definitely a desire to move beyond uniques in geography, not that uniques are not important. My own predilection is to do comparative regional geography, which would mean getting a lot of details on regional arrangements in many places on the earth and then drawing generalizations." [Letter to David A. Skellie, 10 May 1974]

CHAPTER 4   Regional Structure and Arrangement

From 1951 through 1955 Edward Ullman received support for his research from the Office of Naval Research in Washington, D.C. One of his most delightful and provocative articles was a mimeographed report prepared during this time, entitled "Regional Structure and Arrangement" (December 1954), a version of which he read at the interdisciplinary session at the American Sociological Society meeting in Urbana, Illinois. This paper is a result of an unconventional mind probing the conventions of geographic space.

About one hundred years ago Emerson remarked, "Tis a very old strife between those who elect to see identity and those who elect to see discrepancies and it renews itself in Britain."*[1] This strife also "renews itself" in arguments about regions. Regional enthusiasts, particularly those concerned with small areas, are accused of seeing only discrepancies and not working toward broad universals, the goal of science. Universalists, or large region enthusiasts, on the other hand are accused of glossing over discrepancies to produce spurious identities.

As is often the case there is a valid middle ground between these two poles. A central problem in countering the objections of the two extremes is to establish regions of maximum distinctiveness—to set up the largest valid regions for specific purposes which do not average out significant differences. Using a region as large as the whole of the United States to depict population trends, for example, as Donald Bogue has noted, obscures significant facts and conclusions.[2] No single large geographic region of the United States paralleled the national trend from 1940 to 1950. To discover the real reasons for these changes, or for countless other occurrences, it is necessary first to regionalize the data properly in order to provide meaningful material for analysis. On the other hand to confine analysis to a vast number of separate, individual places, although useful for many purposes, overlooks some meaningful generalizations for larger areas.

The size and type of region to be set up depends on the criteria used and the purpose intended as many geographers and other scholars have noted. Data therefore should not be forced, in many cases, into a rigid set of preconceived regions, however well intentioned and thought out, or even worse, into conventional statistical units. Nor for every investigation should arbitrary new regions be coined at random, which obscure correlations and causal connections with other distributions. Many arbitrary regions merely enclose space and resemble trash cans; if they demonstrate useful distinction, it is merely on the level of distinguishing between trash cans and garbage cans. Intelligent recognition of regional

1. Ralph Waldo Emerson, *English Traits* (c. 1855). Reprinted in "Literature," in *American Essays*, ed. Charles B. Shaw, p. 11.

2. Donald J. Bogue, "The Geography of Recent Population Trends in the United States," *Annals, Association of American Geographers* 44 (1954):134.

*Revision of a paper presented to the American Sociological Society, Urbana, Illinois, 10 September 1954. Thanks are due the Office of Naval Research for financial support.

types and skill in varying areal arrangements often makes possible deeper penetration and understanding.

As an attempt to aid in this penetration the remainder of this paper will define two broad aspects of the spatial component of regions: (1) the two irreducible and contrasting types of regions (homogeneous and nodal), each a result of a quite different process; and (2) the problem of attaining regions of maximum distinctiveness within a framework of biased statistical divisions.

HOMOGENEOUS AND NODAL REGIONS

Of first importance in analyzing the spatial aspects of regionalization is the two-fold division of regional types into: (1) homogeneous or uniform regions, and (2) nodal or functional regions. This might be compared to the difference between anatomy and physiology, or morphology and hydrology, although no further organic or physical comparison is intended. A homogeneous or uniform region is simply one characterized by similarity in one or more criteria such as crops, type of agriculture, or kind of population. Nodal or functional regions are defined on the basis of a quite different process and refer to the connections in area, most notably, for larger regions, the spheres of influence or trading areas of cities. Only by chance do these two types of regions coincide. Thus there is no logical reason for the South to have a capital nor any rationale for Chicago's or any other city's hinterland to be composed of even remotely similar characteristics.

In point of fact trade and interaction often tend to take place between unlike regions. Sometimes this may create a node between the contrasting areas as in the gateway cities along the Missouri and Ohio Rivers in the United States serving as intermediaries between the East and West and North and South. Many of the capital functions of the Great Plains thus are performed by cities on its border such as Minneapolis, St. Paul, Omaha, or Kansas City, and for the South by St. Louis, Louisville, Cincinnati, Washington, or Baltimore. Two of the most distinctive homogeneous regions of the United States, therefore—the Great Plains and the South—have their capitals outside their borders, although the South is rapidly altering with the growth of its own capitals of Atlanta, Birmingham, Memphis, and Dallas.

The distinction between homogeneous and nodal applies to both large and small scales. A local valley, for example, may be characteristically composed of two distinctive homogeneous regions—the fertile, level, populated valley

"On the first page you will note that I did borrow your figure of speech comparing arbitrary regions to trash cans. I think this is the only thing I have borrowed without credit, although some other material is relatively well known to all geographers." [Letter to Derwent Whittlesey, 17 September 1954]

". . . is Los Angeles or New York the center of innovation for the United States? My impression is that it is more Los Angeles. Are the countrysides more different from the cities or vice versa? Today in the United States, I think a good case could be made for the small town in the countryside as a part of the national culture because all they listen to and look at is television, where in the city there are other alternatives, although usually not taken advantage of. The trouble with getting specific is that it is hard to get measures, and without measures, all we have is assertions." [Letter to Lutz Holzner, 2 November 1967]

3. The following four footnotes refer to publications recognizing the distinction, with a remarkable number coming out since 1953. Prior to that time the remarks by Platt and Jones had been virtually ignored, although I wish to acknowledge learning clearly of the distinction from Platt in 1935, and in turn passing it on to Whittlesey who coined the new terms "uniform" and "nodal." Many others have noted the distinction in passing and even written at length on their characteristics, although not sharply noting that these two types represent an irreducible and often contrasting duality. Note especially the full and admirable treatment in Howard Odum and Harry Moore, *American Regionalism* (New York, 1938), and other works by Odum and others noted subsequently.

4. Derwent Whittlesey, et al., "The Regional Concept and the Regional Method," in *American Geography: Inventory and Prospect*, ed. P. E. James and C. F. Jones (Syracuse: Syracuse University Press, 1954), pp. 36–44. See also E. L. Ullmán, "Human Geography and Area Research," *Annals, Association of American Geographers* 43 (1953):58–59, and Edward A. Ackerman, "Regional Research—Emerging Concepts and Techniques in the Field of Geography," *Economic Geography* 29 (1953):192.

5. R. S. Platt, "Field Approach to Regions," *Annals of the Association of American Geographers* 25 (1935):171. R. S. Platt in a statement in *Regional Factors in National Planning* (Washington, 1935), p. 148.

6. Hans Boesch, "Some Major Problems in Geography," *Indian Geographical Journal*, Silver Jubilee volume, 1951, pp. 224–28. G. W. S. Robertson, "The Geographic Region: Form and Function," *Scottish Geographical Magazine* 69 (1953):49–58. Robertson gives particular credit to Hans Carol, "Die Wirtschaftlandschaft und ihre kartographische Darstellung," *Geografica Helvetica* (1946), pp. 246–79 for analyzing the concept but notes that others also had the idea and even used the same terms. Note also the contemporary Italian contribution by Aldo Sestini, "L'organizzazione umana

floor devoted to agriculture, and the infertile, steep, unpopulated valley sides covered by forest. Two more contrasting milieux can scarcely be imagined; yet the valley floor and sides are commonly thought of as one unit, and they often are, functionally. Farms may contain a portion of valley bottom in feed crops and pasture and woodland on the hillsides which are integrated in *daily* operation, one part complementing the other. This fact had to be recognized in establishing the "taking" lines for acquisition of land around new reservoirs in the TVA. Depriving a farm of its lowland made it a non-functioning economic unit. In other cases transhumance (moving of animals from winter valley bottom to summer mountainside pastures) is practiced on a *seasonal* basis. In still other economies where animals are uncommon, as in Japan, the hillsides are intimately tied to the valley bottom by providing firewood and many other complementary products.

Only recently has the distinction between the two types of regions become firmly imbedded in geography in this country and abroad.[3] It should be just as firmly recognized by all disciplines, so that when the term region is used, it will not mean one type to one person and another to someone else, as is the case now. Unfortunately no clearly defined terms have been fixed upon to describe the types. The Regional Committee of the Association of American Geographers suggests the terms "uniform" and "nodal."[4] Previously the terms "homogeneous" and "functional" had been used by some scholars.[5] In Europe at least two scholars from Switzerland and Scotland have recently used the terms "form" and "function" or "formal" and "functional."[6] Unfortunately the adjective "formal" has as first connotation another meaning in English than that implied by the noun "form." Likewise use of the word "uniform" implies either complete uniformity within a region or that all uniform regions are the same, and therefore is not satisfactory to some people. And, finally, homogeneity may be applied to homogeneity of nodal as well as of uniform characteristics! I don't propose to settle this argument nor take any more time on it, but merely refer you to the papers cited. In any event the simultaneous and wellnigh independent recognition and emphasis on the distinction by modern geographers all over the world in the last two years is evidence of the importance they attach to it, although the idea itself is old, going back in unnoticed geographic literature at least to the middle 1930s.[7]

In considering multipurpose nodal and even homogeneous regions, one should recognize the obvious but perhaps nevertheless profound fact that a region exists around each

person and place. Thus someone living on the border between the Great Plains and the Rockies would consider that
neighboring parts of both these areas were in his region. In
other words an underlying characteristic of regions is the
multitude of overlapping regions centered around individuals, one reflection of the very real friction of distance. (A
homely example is provided by a Pole in Chicago who was
asked to delimit his neighborhood. He answered that his
neighborhood was as far as he was "gossiped about.")

Two corollary tendencies also ensue from this fact: (1) a
dense core of population tends to produce a more definite
region than a sparsely settled periphery inasmuch as the
overlapping boundaries around this core tend to coincide for
many individuals instead of for one or a few; and (2) conversely, in zones of sparse population or barriers to movement, a similar piling up of boundaries tends to occur because
of the relative lack of connection across the blank spot or
barrier.

These tendencies are well shown in Latin America (as well
as in other parts of the world) where many of the cores of
population have grown up as separate countries separated by
blank spaces, in many cases not even traversed by roads of
any sort.[8]

"I am still very much interested in regions. Perhaps a useful concept to point out to you would be the difference between a uniform (homogeneous) region, i.e., a region with essentially similar environmental and cultural conditions, such as the South or the Corn Belt area as contrasted with a functional (or focal) region — one which is essentially the tributary area of a large city. Often these two types of regions do not coincide and people do not realize what type they have in mind. . . . One of the striking features about the United States is that the grain of the country is north-south and yet our ties are more east-west." [Letter to B. A. Botkin, 19 February 1952]

PROBLEMS OF ATTAINING REGIONS
OF MAXIMUM DISTINCTIVENESS

For the geographer, or at least this geographer, regions are
merely working tools to achieve some end related to spatial
analysis, the problem being simply to get the best com-
promise in a region to serve the descriptive purpose. Outside
the obvious problem of being sure what phenomenon or
phenomena the region portrays, the main problem, as noted
before, is to get the maximum size area which does *not* average out significant differences. This is a problem on all scales
of standard statistical divisions applying to city blocks, census
tracts, counties, states, and groups of states.

The Smallest Statistical Units: Blocks and Census Tracts

For example, the smallest "regions" for which the census
bureau furnishes data, city blocks, are not inherently good
"regions." Land use and housing quality change in the middle of blocks, not on the street sides, so that the block as a
unit may present a meaningless average. The half block (or in
a few cases the quarter block) is the proper minimum-size

dello spazio terrestre," *Rivista geografica Italiana* 59 (1952):73–92. (This does not as sharply differentiate form and function as the others cited earlier but does an excellent synthesis on the functional type.)

7. Cf. W. D. Jones "Procedures for Investigating Human Occupance of a Region," *Annals of the Association of American Geographers* 24 (1934):107.

8. Cf. R. S. Platt, *Latin America: Countrysides and United Regions* (New York: McGraw-Hill, 1942), p. 84, map on p. 529.

unit to show up maximum differentiation, the goal of a proper regionalization. This same criticism can be applied to other scales; the statistical or other units we use as a matter of course may not be the best building blocks for setting up even larger regions. Census tracts in cities also tend to be bounded by main streets; they are thus poor units for plotting retail trade data, since stores traditionally cluster along both sides of main streets or at intersections.

"Long Lot Units": Many Counties and Most States

Still larger areas may exhibit a bias toward a meaningless average for still other reasons. This bias applies to all "homogeneous" divisions made on the so-called long lot basis, a type of division employed in America initially and most characteristically in French Canada and Louisiana in laying out lots at right angles to a waterway or focus. The same principle was applied to counties in Quebec along the St. Lawrence, in Southern California to counties extending from oases near the coast out into empty desert, or in Oregon to counties extending from the middle of the Willamette Valley into the foothills and mountains. You have all seen the absurd but standard maps of population density or other phenomena by counties exhibiting this error. Counties in much of the Mid-West are not biased particularly in this direction simply because so much of the country does not differ radically within a few miles; counties therefore furnish fairly good building blocks. In New England counties are poor units for another reason; they are larger than in the rest of the United States, a grave defect in view of the small scale of differences characteristic of New England. Towns must be used as building blocks in New England and are so employed by the Census in delimiting metropolitan areas, whereas counties are used in the rest of the United States.

When the states are considered as units, the systematic bias toward the meaningless average is more distressing. The bias is not as extreme as for long lot counties, but is more widespread. Along the Atlantic and Pacific Coasts of the United States most states extend across the grain of the country and into the interior; in the Middle West the north central states of Minnesota, Wisconsin, and Michigan extend from productive farmland northward into sandy or rocky, forested wilderness; even the middle states of Illinois, Indiana, and Ohio extend southward from good areas to poorer, hilly, non-glaciated territory; the Great Plains states have their eastern margins in humid, productive areas, their western margins in semiarid,

sparsely populated regions. On an area basis this long lot
tendency of the states toward the meaningless average is
even more apparent since the smallest states, Rhode Island,
Connecticut, Delaware, etc., which have the least long lot
characteristics, obviously cover the least area. Close to 90 percent of the area of the United States is estimated as subject to
this systematic bias toward the meaningless average. The
facts cited on which this conclusion is based are so well
known that it is surprising that this bias has never been noted
before to my knowledge. Virtually only Iowa of the large
states exhibits a high degree of homogeneity. No wonder it
consistently comes out at the top in many state indicators
ranging from percent crop land through per capita number of
telephones, to lowest illiteracy.

The comments above apply to measures of uniformity or
homogeneity but states today are even worse for nodality, as
witness the metropolises with which you are all familiar
which lie near or astride many state lines.

States have also been objected to on other grounds as witness the statement of Raymond D. Thomas:

. . . state boundaries do not mean much in useful social science
research. Our states are historical accidents. . . . [A] county-
indices' divisional map . . . in certain particulars is to be preferred
to the "state line" map. . . . The larger the area the more difficult is
the problem of discovering regional homogeneity. More and more I
am leaning toward the smaller region, both for purposes of useful
research and for possibilities of regional action—planning legisla-
tion, institutional articulation, etc. It may be that in the social sciences our research should be concentrated more than it has in the
past with smaller units and areas.[9]

This is an admirable statement, but one does not necessarily need to take the gloomy view that regional analysis must
be concerned with areas smaller than states in order to rectify
all the misapprehensions that state units create. In many
cases if it is simply recognized that most states have a persistent bias toward a meaningless average, one can often work
reliably with areas even larger than states if the state lines are
ignored as boundaries of regional data.

All of the above is not to say that states cannot be used at
all. The sheer volume of work to get data for small units, as
well as the total absence for areas smaller than states for much
data, compels use of state areas in many cases. Likewise, the
states themselves create some internal homogeneity and
nodality and would be extremely difficult to abolish or
change, as has been indicated elsewhere.[10] However, a truer
and richer regionalization will result from using smaller or
different units as building blocks. Particularly welcome,

9. Raymond D. Thomas quoted in
Howard W. Odum, *Southern Regions
of the United States* (Chapel Hill,
N.C., 1936), p. 536.

10. Edward L. Ullman, "Political
Geography in the Pacific Northwest," *Scottish Geographical Magazine*
54 (1938):236–39; and "The Eastern
Rhode Island-Massachusetts Boundary Zone," *Geographical Review* 29
(1939):291–302.

therefore, is the work of Donald Bogue and others resulting in the Census Bureau creating State Economic Areas and particularly Economic Subregions, which divide the country into a greater number of units transcending state lines where necessary.[11]

Large Units: Groups-of-States Regions

Some of the standard multi-state regions of the United States are also misleading areas for some of the commonest purposes for which they are used, even though the states can be retained as building blocks. Two examples will illustrate this, each relating to homogeneity and nodality, and to creating regions of maximum areal differentiation.

New England commonly includes Connecticut, an area so close to New York City that part of it is even a suburb. Because it lies mainly in New York's immediate hinterland rather than Boston's it is the least "New England" state in many characteristics. This applies particularly to the important characteristic of rate of growth. From 1940 to 1953 all the New England states combined, except Connecticut, increased in population 12 percent as contrasted to the United States average of more than 20 percent; but Connecticut increased by more than 26 percent, more than twice the rate of the rest of New England.[12] Combining Connecticut with the rest of New England results in an increase of about 15 percent for all New England. Northernmost New England fares even worse in relation to the national average: the states of Maine, New Hampshire, and Vermont increasing only about 7 percent in the thirteen-year period. Connecticut's better showing than the rest of New England is attributable in part to: (1) its having two growth industries, metal fabricating, and insurance, particularly in Hartford; (2) the absence of textiles, a declining industry; and (3) suburban overflow from New York City in the southwest corner of the state. How much the growth in the rest of Connecticut reflects the stimulus of greater proximity to New York and its overflowing influences is unknown, although undoubtedly of some importance. Even without this factor, Connecticut, in growth, and even more so in *trend* of increase, is so radically different from the rest of New England as to deserve separation in this critical index. The nodal region focusing on New York should be recognized for many purposes.[13] Most of New England is worse off than it thinks it is.

Still another example provides an illustration of overemphasis on a nodal factor and a consequent misrepresenta-

11. Donald J. Bogue, *Economic Subregions of the United States*, Bureau of the Census (Washington, D.C.: Department of Commerce and Bureau of Agricultural Economics, publication no. 19, June 1953); and *State Economic Areas* (Washington, D.C.: Bureau of the Census, 1951).

12. Calculated from *Census of Population, 1950* and *Current Population Reports*, series P-25 (Washington, D.C.: Bureau of the Census).

13. Cf. the recognition of this even as early as 1930 cited by John K. Wright in "The Changing Geography of New England" in *New England's Prospect: 1933* (New York), p. 473.

tion of area for some fundamental purposes. In the six-fold regionalization of the United States by Odum, a convincing case is made for including Nevada with the Pacific Coast states.[14] Various indices on a per capita basis particularly show this, as well as ties to California. On an area basis (in other words how the region looks on a map), however, this allocation is misleading. The overwhelming bulk of Nevada's population is concentrated in the vicinity of the two gambling suburbs of San Francisco and Los Angeles, Reno and Las Vegas, respectively, on the western margin of the state. The rest of Nevada is relatively empty and resembles more the intermountain, arid, western region or regions, if statistics are plotted according to area rather than per capita. The 160,000 people in Nevada are so few that they will have little effect on any larger regional indices except those relating to area. Therefore, on the basis of the critical factors of large empty area Nevada should be classed with the intermountain region which is most like it in this one respect. The three Pacific Coast states already have enormous natural contrasts with their combination of humid coast, high mountains, and arid interiors, which work toward a meaningless average. However, because of the strength of their intra-state cross grain or nodal ties (as McKinley and Stewart have admirably demonstrated for the Pacific Northwest), one can argue that they should be regarded as a unit for many purposes.[15] For nodal purposes the same could be said for Nevada, because of its intimate ties with California, but not for some critical homogeneous purposes.

THE GRAND DIVISIONS OF THE UNITED STATES:
TWO WAYS OF DIVIDING THE COUNTRY
INTO TWO CONTRASTING REGIONS

As an illustration of achieving the largest valid region, two sets of grand divisions each dividing the United States into two parts will be proposed. These regions must ignore state lines to achieve maximum impact. The first is based primarily on a fundamental difference in natural and related cultural characteristics between the two parts, and the second, primarily on the basis of a difference in importance and function.

A fundamental two-fold division of the United States into two almost equal parts would divide the country into the humid East and the essentially semiarid West, over most of which crops cannot be grown without irrigation. The western half of the country contains only about twenty million persons, only about one-seventh of the population of the humid

14. Odum and Moore, *American Regionalism*, p. 274.

15. *Regional Planning, Part I: Pacific Northwest* (Washington, D.C.: National Resources Committee, 1936), chap. 3 by Charles McKinley and Blair Stewart. See also Edward L. Ullman, "Rivers as Regional Bonds; The Columbia-Snake Example," *Geographical Review* 41 (1951):210–25.

half. The boundary between the two regions would lie out in the Great Plains in the western parts of the Dakotas, Nebraska, Kansas, and Texas. Population density, railways, roads, and a host of other features markedly thin out in this area in the vicinity of the 100th Meridian at about the sixteen-to twenty-inch rainfall line. Such a division is superior for most general purposes to the common, arbitrary division along the Mississippi.

Note that the boundary lies out in the middle of the flat, easily traversed Great Plains which themselves constitute a distinctive homogeneous region. In this division, however, the Great Plains act as a boundary, which, like almost every regional boundary, is a zone, not a sharp line. Nature appears to abhor a sharp boundary as much as a vacuum. Because of this zonal quality of boundaries, state lines in some cases are not hopelessly impossible as boundaries, although the argument often is compelling not to use them in order that the dividing line be in the middle rather than on the edge of the boundary zone.

Another fundamental, large scale division of the United States into about two parts would separate out the industrial belt of the country as one region and the rest of the country, the South and the West, as another (see map 5.1 in chapter 5, p. 82). The industrial belt extending from about Portland, Maine, on the East to Milwaukee on the West and then south to St. Louis and across central Illinois, Indiana, and Ohio to Baltimore is a tiny region containing less than 8 percent of the United States area, but about 68 percent of the industrial employment and 52 percent of the income (area I on map 5.1).[16] It also serves as the great market for the country; it is the focus for the flows of commodities and is the basis for the alignment of transport routes in the whole country. It even contains about 70 percent of the persons listed in *Who's Who*.[17] The rest of the country might be considered almost a raw material province managed by absentee ownership from this industrial core.[18] If the whole of the United States is considered one nodal region, as it is, the industrial belt then is the core or heart of this functional region.

If the contiguous parts of the extraordinarily productive corn belt, and the edges of the manufacturing belt (mainly Iowa, southern Minnesota, easternmost Nebraska and Kansas, and northern Missouri—area III on map 5.1) are added to this region, the combined area is increased to about 15 percent of the United States, its income to 59 percent, and its proportion of industrial employment to more than 73 percent. This procedure combines the overlapping industrial and ag-

16. Population calculated from county data in *Census of Population, 1950;* manufacturing employment from county data in *Census of Manufactures, 1947;* income from county figures in U.S. Bureau of the Census, *County and City Data Book, 1949* (Washington, D.C.: U.S. Government Printing Office, 1952).

17. Calculated from locality data in "Geographical Index-Non Current Listings-Necrology, etc.," supplement to *Who's Who in America,* vol. 28, 1954–55, Chicago.

18. Cf. *The Southwest.* A Report by the Committee on the Southwest Economy to the President's Council of Economic Advisers (Albuquerque: University of New Mexico, June 1954), especially chap. 11, "The Southwest in the Nation—Some Interregional Relations."

ricultural hearts of America into one dominant region con-
stituting a regional node for all America, with the country's
flows focusing on it to a remarkable degree. The only market
area of similar intensity, but minute in size by comparison, is
the southern half of California.[19] (Adding the contiguous part
of southern Ontario—area II on map 5.1—results in a single
combined region—areas I, II, and III—with only 3.7 percent
of total United States and Canadian territory containing more
than 70 percent of the industrial employment.)

These two sets of grand divisions of the United States aid in
understanding the grand synthesis of the United States better
than any other regionalization, although other scales are
necessary to provide still further details. The grand divisions
have been most useful in giving Europeans a rudimentary
understanding of America; the first division helps them understand why the overall population density of the United
States is not apt to approach that of humid Europe; the second division even surprises most Americans who either do
not realize that Chicago is less than a third of the way across
the continent from New York or who assume that the United
States market is equally nation-wide. For areas outside the
northeastern quadrant, remote location is one of their most
persistent handicaps whether related to access to markets,
remoteness from decision making centers, possibilities for
mass production, or availability of technical and other
specialized skills, services, management, and research.

Still other grand divisions could be made for other purposes. The most common is separation of the South, traditionally the home of perhaps the most distinctively different
culture in the country. This is so well known that it will not be
discussed here. These three divisions bring us back to the
most common and, for many purposes valid, grand regionalization of the United States—the North, the South, and
the West, although the details of their boundaries may vary
from the common conception.

THE FUTURE OF REGIONS

As a concluding note let us consider the future of regions.
Are regional differences between parts of the United States
increasing or decreasing, especially when considered from an
overall viewpoint of perhaps the feelings and characteristics
of the regional populations? The differences apparently are
clearly lessening—people in all parts of the country appear to
be getting more alike in habits, customs and income. Several
factors, many of them well known, help explain this, includ-

19. Cf. forthcoming article by
Chauncy D. Harris, "The Market as
a Factor in American Geography,"
*Annals of the Association of American
Geographers,* 1954 or 1955.

ing decline in immigration and second generation absorption of foreign culture groups into American life, increased communication and mobility, and finally change in working conditions and occupational structure. Let us examine the varying impact of the last two factors.

Increase in communication and transportation, because of improved technology, enables rapid, cheap, and widespread interchange of uniform ideas, entertainment, and products. This has both good and bad results. On the credit side, for example, standards are brought up across the nation, often with the aid of the federal government, or in other cases, simply from increased interaction and knowledge; attempts at educational equalization are seriously made, good roads run everywhere, and it is even possible to get a decent meal anywhere in the United States, although one should not count too heavily on good food, good education, or many other "good" features in vast regions far from metropolises in the intermountain West or the South. On the deficit side, the local newspaper in many sizeable cities has degenerated to printing mainly canned news and features provided by national syndicates, and one hunts in vain for good interpretations of the local region, although international news may be provided. This is merely one example of the impoverishment of regional originality.

In one sense, however, easy transportation should result in greater differentiation, since it enables regions to specialize in the activities each can do best, based on difference in factors of production related to the differing physical and cultural environments or simply to economies of scale. This has happened to a notable degree in the United States with the development of large, specialized argicultural and industrial belts. The change, however, has been primarily one of scale. Within the wheat belts, for example, there is less internal, subregional differentiation than would be required if these areas had to produce most of their subsistence needs.

The tendency toward specialization just noted may be counterbalanced by the change in occupational structure within the country. Formerly a large part of the population was engaged in basic agricultural and extractive pursuits which varied more or less directly with the variation in environment of the regions. Today the percentage of population engaged in these pursuits has declined drastically with mechanization in agriculture and in some other extractive pursuits, and particularly with the rise of services in our increasingly wealthy and specialized economy. The percentage of population employed in Colin Clark's categories of primary activity (agriculture, forestry, and fishing), secondary (manufacturing, mining, and construction), and tertiary (trade and services) changed as follows from 1850 to 1950 in the United States.[20]

	Primary %	Secondary %	Tertiary %
1850	65	18	18
1920	27	33	40
1950	17	34	49

Almost half the population is now in tertiary activities, services most of which cannot be shipped from place to place. A clerk or a storekeeper is essentially the same (or varies intraregionally much the same) anywhere in the country, in contrast to the difference between the life and outlook of a subsistence farmer in a poor soil region and a rich farmer in a good area, or between a corn farmer and a fruit grower or tobacco raiser. Many factory workers also tend to be regionally similar, although specialization is greater regionally than in services. As a result, men tend to earn their living less and less on regionally specialized activities, although it is true that some regions are predominantly rural and others industrial, and contrasts will always remain. The farmer of Iowa, however, regularly now shops in supermarkets, even for groceries, just as does his fellow factory or service worker in New England.

The foregoing applies only to working hours, but the vastly increased leisure time of Americans may work in the opposite direction to the extent that people do not spend their leisure time in watching canned movies and television or reading national magazines or local newspapers with no local color. Leisure activities carried on outdoors, whether in the mountains, at the seashore, or on the patio when the sun shines, may vary more regionally. If none (or only a few) of these features exist in a region, they are not factors. Therefore regions with the greatest number of natural amenities of climate and/or scenery and outdoor attractions may evoke a different pattern of leisure activities. Parts of the South and the West may thus develop different enthusiasms (children may play outdoors instead of reading books, etc.), and evidence seems to indicate that they do, although fish are planted in streams and lakes everywhere, and California patios and sportswear seem to penetrate the whole country, even though unusable in winter.

As a final remark, even though regional differences appear to be waning, it has been demonstrated that the statistical devices and areas in common use today do not portray as much regional differentiation as actually exists. To show up

20. Cf. Edward L. Ullman, "Amenities as a Factor in Regional Growth," *Geographical Review* 44 (1954):119–32.

the maximum, significant variation for specific single or large, interrelated groups of multiple purposes it is necessary to vary the areas used as well as the criteria. Arbitrary, new and thoughtless regional areas should not be coined at random, but equally bad would be to freeze the areal variable and thus blot out deeper penetration and understanding.

In the future new bases and arrangements of regional structure as a part of changing spatial relations will develop, and call for new measurement, analysis, and interpretation. The cotton belt has moved west as far as California and stock-raising eastward to the Old South; the corn belt is moving north into Minnesota; the steel industry is expanding in the populous Atlantic market, along with economies in use of coal for fuel, dwindling of Mesabi ore reserves, development of Venezuelan and Labrador iron deposits, and abolition of Pittsburgh plus; oil and natural gas are piped everywhere; people are moving by auto and radically altering traditional interaction patterns; southern workers are moving north to Detroit and it seems that everyone who can goes west to California or south to Florida. Gross features of the present regional structure will persist as far as the foreseeable future, but evolving changes will require new analysis; in the unforeseeable future with possibilities of solar energy or desalting of sea water, who can tell what will happen? There will always be regions but they will not be the same.

Regional Development   CHAPTER 5
and the Geography of Concentration

The concentration of development, especially industrial, in a few parts of the world, notably the United States and Western Europe, is widely recognized. The contrast in development is just as pronounced in the internal structure of countries, whether it be Northern Italy vs. Southern Italy, or Central Japan vs. Northern Japan. Equally notable is the disparity within the United States, where, in the northeast, about 7 percent of the area has about 70 percent of the nation's industrial employment; the rest of the country fights for the remainder in a manner not unlike a pack of hungry dogs fighting over a dry bone. The major policy and research problem therefore is created by the other side of the coin—underdevelopment elsewhere.

Our main concern will be the extent of the concentration, particularly in the United States, and its effect on development of the fringe areas. As to reasons for the concentration it must be conceded that most of the core areas have or had remarkably better natural endowments particularly in productive plains areas than the fringes. Even economists would concede that in Canada, for example, the contrast between southern Ontario and the vast cold northern stretches and the rocky Laurentian Shield provides necessary and sufficient conditions for relative non-development in the latter sections. Australia, Switzerland, the United States, and many other countries provide somewhat similar examples of extreme natural differences.

In relating non-development to lack of resources one explains the phenomenon in terms of itself, but such is the nature of resources; they are not resources unless man's technology can use them. It was a fortunate coincidence that the Industrial Revolution, in that important phase of its evolution which used coal for generation of steam and manufacture of steel, occurred close to the coal fields of western Europe and eastern United States. In this case precisely what is cause and what is effect is difficult to determine.

Equally important, once the concentration gets started, is the self-generating momentum of the concentration itself. The concentration becomes the important geographical fact.[1] A host of complementary activities and services is established, each helping the other in pyramiding the productive

This paper was published in the Papers and Proceedings of the Regional Science Association *(1958), pp. 179–98, and has been reprinted in several places, including the* Bolletino della Società geografica italiana 12 *(1959):319–44. Its theme is that great regional disparity exists within areas, particularly within nations and states that are often treated analytically as homogeneous, of such a serious nature that data may constitute "meaningless averages." In contrast to the physical regions of the United States from which divisions commonly were made in accordance with homogeneous features—for example, the corn belt—Ullman suggests divisions based on the concentration of development, which recognize a core-central-periphery hierarchy. Thus, he introduces the concept of a core, or market area, as an area which largely controls other regions. He first explored this theme in "Regional Structure and Arrangement" (chap. 4), written in 1954.*

1. Cf. C. D. Harris, "The Market as a Factor in the Localization of Industry in the United States," *Annals of the Association of American Geographers* 44 (1954):315. See also Gunnar Myrdal, *Rich Lands and Poor: The Road to World Prosperity* (New York: Harper, 1957), chap. 3.

"In fact no where else in the world, with the possible exception of China, do I from a cursory glance find gateway cities such as those in Canada and the United States between the industrial belt and the rest of the country. Of course I exclude ports which are gateways anywhere. Why this should be so I don't know, I doubt that it has anything to do with capitalism or even the timing of development." [Letter to Chauncy D. Harris, 11 August 1970]

2. Institute of Pacific Relations, *The Development of Upland Areas in the Far East* (New York: 1951); Pierre Gourou, "The Quality of Land Use of Tropical Cultivators," in *Man's Role in Changing the Face of the Earth*, ed. William Thomas (Chicago: University of Chicago Press, 1956), pp. 336–49.

3. U.S. population calculated from county data in *Census of Population, 1950;* manufacturing employment from county data in *Census of Manufactures, and City Data Book, 1949* (Washington, D.C.: U.S. Government Printing Office, 1952). Canadian data from *Census of Canada: 1951*, vol. 4, "Labor Force, Occupations and Manufactures" (Dominion Bureau of Statistics, Ottawa). Note also C. D. Harris' calculation that the belt contains 50 percent of U.S. retail sales ("The Market as a Factor in the Localization of Industry," p. 319).

4. Calculated from locality data in "Geographical Index-Non Current Listings-Necrology, etc.," supplement to *Who's Who in America*, vol. 28, 1954–55 (Chicago).

5. Cf. Edward L. Ullman, *American Commodity Flow* (Seattle: University of Washington Press, 1957).

process; the largest market in the country is created, in which transport costs dictate location of much industry if national distribution is desired to take advantage of scale economies. For the fringe areas to develop in the face of this formidable competition poses an almost insuperable obstacle.

This is not entirely correct, since the market of the core area is also generally the principal market for the corner area. However, if every part of the world or of a country were equally developed or undeveloped, by definition there would be no underdeveloped areas. By the same token if undeveloped areas develop at an even slower rate than developed areas, as is often the case, they remain relatively undeveloped. The contrast poses the problem.

Even in most non-industrialized countries there is a remarkable concentration of settlement in one portion; this concentration cannot be explained solely by resource endowment. The better areas are crowded beyond reasonable capacity and the remoter areas, many with reasonably good natural endowments, are relatively empty. Reasons for this are obscure, but probably include difficulty of shifting lowland farming techniques to uplands or vice-versa,[2] lack of access roads, and generally a reluctance, apparently, to move away from neighbors and the amenities associated even with a modest level of social overhead.

The problem of concentration thus is world-wide. Let us now examine the American situation in detail.

CORE AREA OF AMERICA

Map 5.1 shows the concentration area in the United States, about 7 percent of total U.S. area, generally called the industrial belt, since in addition to concentration of population (43 percent of U.S. total) and income (52 percent of U.S. total), it is the center of industry with almost 70 percent of the U.S. total.[3] It has similar or even higher percentages of still other activities; to name but one example, about 70 percent of those listed in *Who's Who*.[4] Altogether this is a remarkable concentration, although, as noted earlier, characteristic of the internal areal structure of most countries of the world. Map 5.2 (railroad traffic) also shows the transport net focusing remarkably on the industrial belt; the same would be true of natural gas and petroleum pipe lines. The core area clearly aligns the major flows in the American economy.[5] Map 5.3, showing shipments of animals and products from the state of Iowa, the leading animal producer, indicates the shipment of the products to the industrial belt and secondarily to California, the southern half of which has, especially since the war, developed into a subsidiary and smaller industrial concentration.[6]

INADEQUACY OF STATE DATA
TO INDICATE DEGREE OF CONCENTRATION

The map (5.1) showing the industrial belt was constructed on a county basis. In this manner it was possible to show the really significant concentration better than by using states as building blocks.[7] Furthermore in the process of making the map it was discovered that most of the states of the United States are set up on a "long lot" basis—one similar to the classic examples in French Louisiana and Quebec where land holdings are laid down at right angles to the major rivers or foci. The same pattern is followed by many minor civil divisions and counties, most notably some in southern California with their nodes in the coastal oasis and their vast extent stretching out into the empty desert. This long lot characteristic is notable for most of the major states of the industrial belt;

6. Map 3 prepared from Interstate Commerce Commission's 1 percent sample of rail traffic reported in *Carload Waybill Analyses*, Washington, D.C. (statements: 4838, October 1948; 492, January 1949; 498, March 1949; 4920, June 1949).

7. Cf. also Harvey S. Perloff, "Problems of Assessing Regional Progress," *Regional Income, Studies in Income and Wealth* (National Bureau of Economic Research) 21 (1957):39.

Map 5.1. Core areas of the United States and Canada: Area I (% U.S.): 7.7% area, 52% income, 70% persons listed in *Who's Who*. Area III (% U.S.): 6.9% area, 7.3% income. Areas I and III combined (% U.S.): 14.6% area, 50.3% population, 59% income, 73.3% industrial employment. Area II (% Canada): 0.4% area, 19.8% population, 33% industrial employment. Areas I and II combined (% U.S. + Canada): 3.7% area, 41.2% population, 65.9% industrial employment. Areas I, II, and III combined (% U.S. + Canada): 6.9% area, 47.7% population, 70.8% industrial employment. Sources: see footnotes 3 and 4.

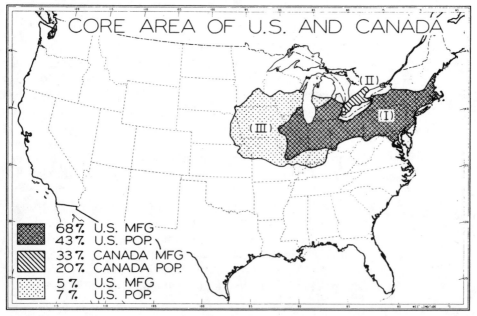

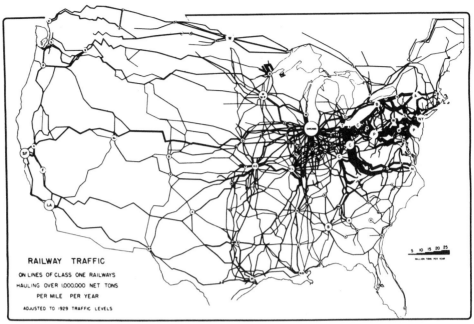

Map 5.2. Railway traffic in the United States and Canada on lines of class one railways hauling over 1,000,000 net tons per mile per year. Width of lines is proportionate to volume (of short tons of 2,000 pounds). Prepared from data copyrighted by H. H. Copeland and Son, New York; Canadian lines added and map adapted by Edward L. Ullman.

Map 5.3. Destination, by states, of animals and products shipped out of Iowa by rail, 1948. Width of lines is proportionate to volume (of short tons of 2,000 pounds).

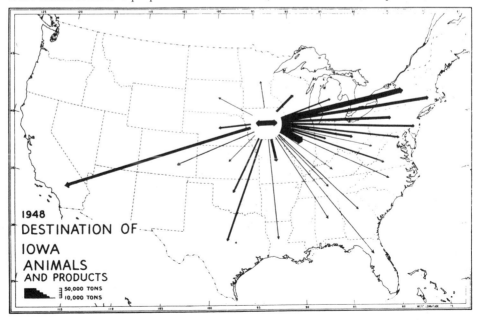

1948
DESTINATION OF
IOWA
ANIMALS
AND PRODUCTS

50,000 TONS
10,000 TONS

New York, Michigan, and Wisconsin extend up into the relatively empty Adirondacks or the sterile glacial outwash plains and ice-scoured rocks of the north. Illinois, Indiana, and Ohio in turn extend southward into relatively less fertile, nonglaciated terrain. To the West, the Great Plains states, from Dakota to Texas, exhibit remarkable long lot characteristics extending from their eastern Gateway nodes, in humid climate, into the semiarid reaches of the western high plains. The Canadian Provinces also strikingly exhibit this characteristic with their long north-south extent. As a result, using most state aggregates creates meaningless averages, masking significant differences, and, for our purpose, biasing the results. Using most states and nations to show national averages thus builds in an error, an error for many purposes and places which might almost be more biased than using a world average, which at least has the merit of combining everything, and not just two diametrically opposite aggregates on which extreme long lot political units are based.

ASSEMBLY AND MARKET ADVANTAGES OF CORE AREA

To return to the industrial belt, the reasons for the concentration are similar to those in the rest of the world: a combination of resources, early start, type of settlement, and what is most important for our purposes, a self-perpetuating momentum resulting from a pyramiding of complementary activities and services to produce notable external economies of scale and the largest market in the country. As a result most assembly and selling can be most economically done in this area. As just one example, 98 percent of the members of the American Machine Tool Builders Association, a critical industrial service, are located within the industrial belt.[8] Only four are outside; two in contiguous Southern states and two in southern California. Chauncy Harris has demonstrated quantitatively and graphically the importance of the industrial belt as the market and place to locate to reach the greatest national customer potential. For the West alone the center of gravity is Los Angeles, so large is its population.[9]

Finally, it should be emphasized that the market is becoming increasingly important as an industrial location factor because of increasing economies in consumption of fuels and raw materials and other factors. In 1900 each dollar of raw material yielded $4.20 in finished product; in 1950, $7.80 in constant value dollars.[10] Likewise it appears that economies in the shipment of bulky commodities in volume—coal, oil, ores—are increasing relatively to the handling costs of the

"Articles are what textbooks are written out of. I think it quite appropriate that more credit be given for original scholarly articles than rehashing stuff for a textbook or only for teaching. Scholarly articles are the first contribution to teaching. They reach a wider market and help educate a larger number of students than those just enrolled in classes of the professor." [Letter to Kenneth MacDonald, 6 December 1973]

8. Data furnished by the Association which includes about 200 of the approximately 300 concerns in the business; over two-thirds of the members are west of Pennsylvania.

9. Chauncy D. Harris, "The Market as a Factor in the Localization of Industry," pp. 315–48.

10. Edward S. Mason, *Economic Concentration and the Monopoly Problem* (Cambridge, Mass.: Harvard University Press, 1957), p. 255.

"For years while I was at Harvard, I used to ask honors candidates for the bachelor's degree in geography where they would draw the boundary if they had to divide the United States into two parts. They invariably picked the middle of the Great Plains, not the Rocky Mountains. In other words, as I indicate in the enclosed paper, there are two types of barriers, some of which Hartshorne has termed 'kinetic,' i.e., mountains and similar physical barriers, and the others 'static' divides, areas with sparse population. Obviously on these latter divides is the place where many of your transcontinental rates make some abrupt changes. Here there are fewer people and consequently less competition, *ceteris paribus.*" [Letter to Stuart Daggett, 22 September 1952]

11. Edward L. Ullman, "Amenities as a Factor in Regional Growth," *Geographical Review* 44 (1954):119–32; *Arizona Business and Economic Review*, vol. 3, no. 4, April 1954, and "A New Force in Regional Growth," *Proceedings Western Area Development Conference*, Stanford Research Institute, 17 November 1954, pp. 64–71.

12. Cf. Lloyd Saville, "Sectional Developments in Italy and the United States," *Southern Economic Journal* 23 (1956):39–53.

finished packaged materials, further drawing industry to the market. However, the increasingly lower grade of many non-fuel mineral products may work in the opposite direction, although the main effect is merely to set up, near the ore, concentration plants which employ few, rather than fabricating establishments employing many.

PROSPECTS FOR THE FRINGE AREAS

In contrast to the core areas the prospects for the fringe or corner areas appear rather bleak, since they are remote from the center of the system and the self-generating momentum of the center. Their best hope is to possess some special lure such as the present role of climate of California or Florida, or, in the past, superior trees in the Pacific Northwest. Only by such lures have the corner areas been able to overcome their remoteness from the industrial belt, as I have noted for amenities in papers published elsewhere.[11]

One special point that deserves attention in the peripheral areas of the United States is that only one part, the South, has markedly lower per capita income than the core area. The West (especially the Pacific Coast) has essentially the same or even higher incomes than the core, whereas the lowest income state in the South, Mississippi, has only about one-third the per capita income of the highest in the North, although states not so deep in the South are better off. This is essentially the same pattern as in Italy, where the lowest per capita incomes, in the southern provinces, are only about one-third those in the north.[12] Presumably somewhat similar disparities exist in other, older settled countries, whereas the new West of the United States has migration into it controlled by real or psychic income available, rather than having filled up earlier on a low income level, or something like that (precisely what is not germane to our problem). In terms of national percentage of concentration, however, the West and the South are similar, except that the low per capita incomes of the South provide a special lure insofar as the lower wage rates attract industry and lower living costs favor labor, although not true to the degree that a lower level of social overhead (schools, roads, etc.) is associated with the lower per capita income and this in turn affects development. Likewise a high labor cost area provides more incentives for innovations in labor saving devices and thus stays ahead of the procession to that degree.

Let us now consider the prospects of corner areas on a purely spatial basis. By definition a corner area would have

less area and possibility for large market and concomitant
thresholds for economies of scale than would a central location. This is a factor of some importance. If the 1,400,000
population of British Columbia and the 2,600,000 population
of Washington were added together a larger market and
greater development theoretically would result. In practice
this happens for some goods and services that can cross the
international border, notably the recently finished long distance natural gas and oil pipelines from the North, which the
market of British Columbia alone could not have supported.
For most industrial products, however, the boundary is a
barrier, as absence of heavy rail traffic across the border
shows (see map 5.2). Southern California, and Mexico
through remittances, however, also benefit from the flow of
Mexican agricultural labor across the border. Miami serves
somewhat as an air and resort center for neighboring Latin
America, etc., and all corner locations on the ocean can attempt to develop sea trade, which opens up the markets of
the world to them. Some leakages thus are available. However in actual practice, even without leakages, corner location
need not be crippling because of the possibility of concentrated lures, as at Los Angeles.

A second aspect of the problem is that areas remote from
the main market, and thus unable to compete nationally, do
by the same token have a protected local market. Here the
problem is to develop sufficient economies of scale; southern
California appears to be crossing this threshold, with many
branch plants, although exactly how far this has gone is
difficult to say.[13]

An institutional factor retarding regional development is
the presence of national competition in the United States.
This means that no one company gets all the business for one
product in a region and hence the threshold of scale
economies is not crossed and a local plant is not established.
National competition in many lines forces industry to locate
in the core and thus results in regional monopoly of the industrial belt, the center of gravity of the country. As a corollary, therefore, one might say that much of our antitrust policy has worked for regional monopoly of the industrial belt;
however, I do not mean to imply that this policy has been any
more than a minor factor, or that it is necessarily wrong.

Protection from competition of the industrial belt can also
produce advantages not related to scale economies. For
example, a large aviation company in the Northwest, the only
one in the area, feels that it has greater labor stability than
other centers with many competitors because engineers and

13. Edward L. Ullman, "A New
Force in Regional Growth" (Stanford
Research Institute, no. 17, 1954), p.
70. Cf. also the statement about the
South, ". . . the new production
represents in many instances a filling
in of the local industrial structure,"
in Glenn E. McLaughlin and Stefan
Robock, *Why Industry Moves South*
(Washington, D.C.: National Planning Association, 1949), p. 125.

labor are not constantly shifting from company to company seeking better jobs; consequently, from management's viewpoint the benefits of isolation just about cancel out the drawbacks.

In general terms, however, it seems safe to say that a region in the commercial American economy cannot be expected to grow unless it has some ways of sharing in some aspects of the total national market or supply area—in other words, be nationally commercial or competitive. Something has to bring population into the area before local service can develop to serve it. This is the traditional pattern for development in the United States, after the earliest subsistence era, as Douglass North has pointed out.[14]

One way to visualize this might be to speculate on what would happen if the United States were divided into a series of three to four truly independent countries. How would each unit fare? Each would have to compete on the world market, instead of enjoying free access to the national market, to the extent this is permitted today by transport costs and national competition. So many variables are present in terms of varying resource endowments, scale thresholds, government policies, etc., that it would be difficult to make a definitive reply, but some general logical conclusion can be postulated.

Insight might be gained first by considering Canada. Canada does have branch plants behind her tariff wall, but can sell only non-competitive products freely across the wall into the United States. Canadian per capita income is lower than that of the United States and most of her consumer industrial products are higher priced, presumably because of smaller scale production. The result is a considerably lower real per capita income and, in the past, a net migration toward the United States. If there were no international boundary one might well postulate that Canadian population would be lower, but per capita real income higher, in spite of the fact that Canada would have less of some manufacturing and other activities traditionally associated with high incomes.[15] (However Canada also is a special case; the loss in branch plants would probably be more than compensated for by the spillover of American national industrial belt production into lower Ontario, which is in the heart of the American industrial belt.) Agricultural and competitive raw material prices also would be higher if Canada had free access to the U.S. market, although the terms of trade could be more readily controlled by an independent Canada in the future as U.S. demand for its resources mounts, than would be the case if it were part of the United States. Canada's present refusal to

14. Douglass C. North, "Locational Theory and Regional Economic Growth," *Journal of Political Economy* 63 (1955):243–58.

15. Cf. Stephen B. Jones' discussion of the draining of trade south down the valleys into the United States from British Columbia via branches of the Great Northern Railway and the counter move of the Canadian Pacific in building the Crowsnest Pass line from East to West to divert these flows to all-Canadian routes. ("The Cordilleran Section of the Canada-United States Borderland," *Geographical Journal* 79: 439–50).

permit reservoir construction to store water in British Columbia for generation of power downstream in the United
States lowers Canadian income at present, but may raise it in
the future. Likewise the St. Lawrence Seaway would not now
be under construction if Canada were a part of the United
States. Urban development, metropolitan services, and some
transport within present Canada (outside the special case of
lower Ontario, if it received national manufacturing) would
be lower if it were a part of the United States and prices of
industrial goods considerably lower.

The general principle at work is the old one that trade benefits both partners considered as a unit but not necessarily
each individually, nor the various sectors equally. Thus, we
might say that if the international boundary did not exist, that
population in much of the smaller country, Canada, would be
smaller, but per capita incomes would be higher because of
fewer members and general spillover of American income
levels and prices.[16]

For the reverse process, creation of several individual countries in the United States, the opposite should occur; incomes
would drop all over but development might increase in the
smaller units, but decrease in the present core area. In both
cases equal relative mobility or immobility of all factors of
production (goods, capital, labor, etc.) is assumed.

INNOVATION AND CENTERS OF CONTROL

If the prospects for the fringe areas are bleak, what are their
chances of overcoming this disadvantage by innovation and
local initiative, especially in developing superior footloose industries and activities which can enter the national market
without undue transport penalty? Can or will they build a
better mousetrap? In attempting to predict possibilities here I
shall present two sets of findings, which are related to the
problem but are by no means conclusive. They are new quantitative measures presented for speculation.

Innovation

Map 5.4 shows issuance of patents by states. The industrial
belt, not surprisingly, has 70 percent of the patents issued in
the country. On a per capita basis the industrial belt states
and California rank highest (map 5.5). On a county basis the
concentration would be even sharper, but data are not available.[17] Location of patents, of course, is not conclusive evidence of innovation, ability, or application. They may be used

16. Here we are talking of advanced countries, where Myrdal
says "spread" effects operate. Cf.
other aspects also as treated in chap.
3 of Myrdal, *Rich Lands and Poor.*

17. Sources of the maps are manuscript reports of the U.S. Patent
Office from which a three-year annual average, 1951–54, was computed.

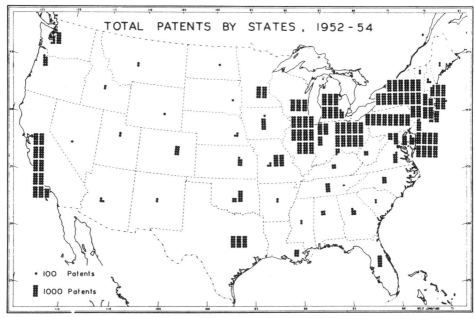

Map 5.4. Distribution of patents in the United States, by states, 1953–54. (For sources see footnote 17.)

Map 5.5. United States population per patent, by states, annual average 1952–54. (For sources see footnote 17.)

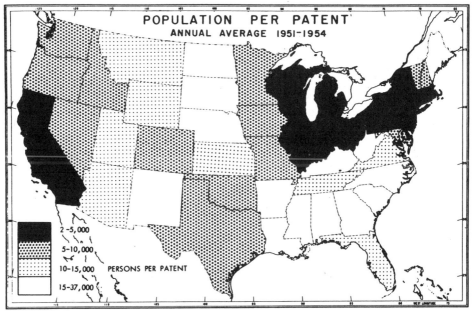

anywhere in the country, especially since many of them are granted to industries with nationwide plants. The situation is not the same today as it used to be when a local invention resulted in a local industry. Nevertheless localization of invention does create many localized industries even today, although exactly how many we do not know.

Turning to some other measures of possible innovation or superior technique, some further maps are presented whose bearing on the problem is probably even less direct, but which may at least be of interest for themselves. They are: location of leading scientists and scholars as represented by (1) members of National Academy of Sciences (top scientific honorary), and (2) directors of Social Science Research Council and directors and delegates of American Council of Learned Societies (humanities and social sciences) (see maps 5.6 and 5.7).[18] Note that the concentration is strikingly similar to industry—in the industrial belt and California, except for relatively greater emphasis on the eastern, older half of the industrial belt. (Perhaps a small part of the concentration in the core area occurs because those close to the core are better known, simply because of more frequent contact).

Other maps prepared by the author but not included here show: (1) location of the members of the American Association of Universities and (2) the size and distribution of the thirty-eight largest university libraries, a better measure of eminence. Again, the industrial belt–California orientation emerges. If the universities were ranked on quality, this orientation would be even more pronounced. A recent survey of twenty-five leading universities by the University of Pennsylvania shows, of the top fifteen ranked, that thirteen are in the industrial belt (or very close) and the other two in California. Distribution of other large libraries or top engineering schools and industrial-economic research institutes would exhibit similar patterns.

A last map in this series (map 5.8) showing book publishing,[19] provides an extreme example of localization; almost all the publishing houses are in New York City. This also reminds us that since the invention of the printing press, ideas can circulate without personal contact, although there is a lag in publication.

How are we to interpret these data? Does the concentration of great minds affect the development of regions? Manifestly, we in the fringe areas today, especially in cities or large institutions, are not as badly off as the Georgia poet Sidney Lanier was in the late nineteenth century when he found in Baltimore what he missed in the South where there was "not

18. Compiled from rosters furnished by National Academy of Sciences, Social Science Research Council, and A.C.L.S.

19. Data for 1954, as reported by *Publisher's Weekly*, 1955.

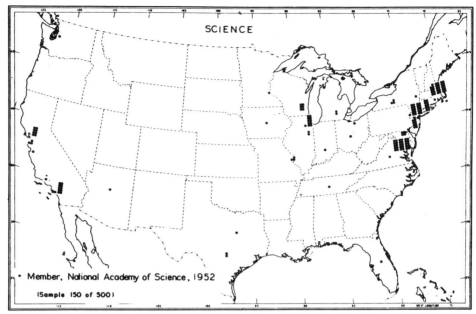

Map 5.6. Distribution of members of the National Academy of Sciences, United States, 1952. (For sources see footnote 18.)

Map 5.7. Distribution of directors of the U.S. Social Science Research Council, 1955, and of directors and delegates of the American Council of Learned Societies, 1954. (For sources see footnote 18.)

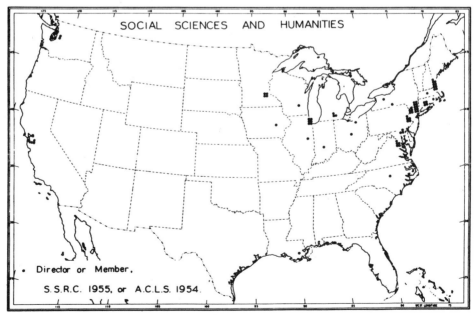

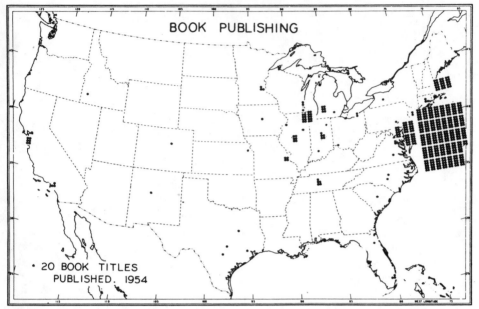

Map 5.8. Distribution of book titles published in the United States, 1954. (For source see footnote 19.)

enough attrition of mind on mind . . . to . . . bring out any sparks from a man."[20] Nevertheless the competition and inspiration from association with superior minds does affect the performance of others. Good also tends to attract good. The concentration of brains therefore has some effect on the future localization of intelligence, a legitimate part of regional development. How much effect this in turn has on the general underpinnings of economic development would be more difficult to determine.[21]

Centers of Decision Making

Maps showing the location of the 100 and 500 largest industrial corporations indicate that about 90 percent measured by value of production (or assets) are in the industrial belt, an even higher concentration than for other activities.[22] (See map 5.9. The map of the 500 largest industries shows essentially the same distribution and is not included here.) This is natural, since most large industries are national in scope. The home office and principal plants however may not be in the same place, but in a majority of cases they are, except for many New York City headquarters. Even in this case, if home office and principal plants are separated, the principal plant is likely to be in the industrial belt. Maps showing the location

20. Van Wyck Brooks and Otto L. Bettman, *Our Literary Heritage* (New York: Dutton, 1956), p. 134.

21. Up to now it has not been the dominant regional influence within all of the United States, as witness the relative decline of New England, in spite of concentration of certain skills and learning there, although strictly industrial research and development are probably not above the U.S. average. However, New England's initial rise, somewhat of a mystery, its subsequent failure to collapse utterly, and its future prospects, may owe something to local ingenuity and invention; research-based industries are developing and apparently will become more important (cf. Richard M. Alt, "Research Based Industries in New England" (Cambridge, Mass.: Arthur D. Little, Inc., 1955), especially pp. 8 and 16–22).

22. Maps compiled from *The Fortune Directory of 500 Largest U.S. Industrial Corporations*, supplement to *Fortune*, July 1957, with assistance from Thomas Directory of Manufactures. U.S. Steel and the Pullman

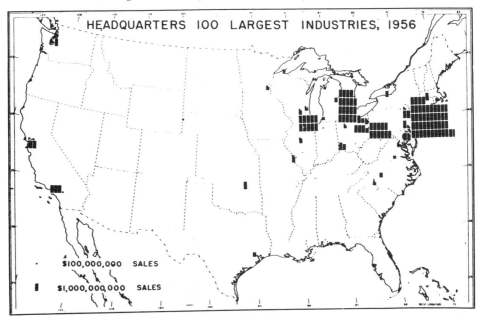

Map 5.9. Distribution of the headquarters of the 100 largest industrial corporations in the United States, measured by dollar sales, 1956. (For sources see footnote 22.)

Map 5.10. Distribution of the headquarters of class one railroads in the United States, by dollar revenue, 1950.

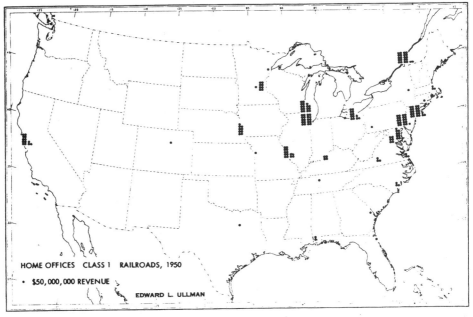

of the 100 largest trade corporations show almost the same concentration in the industrial belt. The map of home offices of railroads (map 5.10), based on revenue, is especially interesting, inasmuch as it shows location on the margins of the industrial belt, at gateways to the rest of the country (except for Cleveland, a readily explainable special case based on real estate speculation). Major exceptions are port gateways elsewhere in the country.[23]

What does this concentration mean? As in patents, a decision to start something new or expand does not necessarily result in expansion at the headquarters of the company. Often, however, in the absence of compelling reasons, or even without reason, the expansion occurs at or close to headquarters. As one executive of a large company remarked to me, "We would be less than human if we did not favor our own location." In any case the research and innovation center is apt to be near home base; furthermore, competing or complementary innovators are regularly spun off from the parent; they establish their own plants, generally on a shoestring, and therefore naturally in the same town. In periods of contraction a company would also be more disposed to close out branch plants than headquarters, as many communities learned to their sorrow during the depression. Exceptions might occur where the branch plant is more modern, but generally fringe areas might be compared to minority groups: "first fired, last hired."

CONCLUSION

Concentration within countries is the rule. This fact may signal the operation of a general localization principle in man's use of the earth: initial location advantages at a critical stage of change become magnified in the course of development.[24] Geographical differentiation starts out as a matter of homeopathic doses of mild concentration and winds up as a system of massive localization based on a wide range of internal and external economies of scale. This direction continues until some radical new change is precipitated and indeed may persist, even though new competitors arise elsewhere, through several such changes including cataclysmic wars or development of new energy sources, as in the shift from wood to coal to petroleum and, in the future perhaps, even to nuclear energy. The really significant geography thus becomes the latter concentration of man himself, his works, his institutions, and his inventive momentum.

This concentration in practice takes the form of an area of

Company, for example, were credited to Pittsburgh and Chicago respectively, since these cities are their effective headquarters, even though *Fortune* lists New York as the legal headquarters.

23. Cf. Edward L. Ullman, *American Commodity Flow*, chap. 2.

24. For a remarkably similar concept note Myrdal, *Rich Lands and Poor*, pp. 26–29. I discovered these statements after writing this paper. They read, in part: ". . . play of the forces in the market normally tends to increase, rather than to decrease, the inequalities between regions," and "occasionally these favored localities and regions offer particularly good natural conditions for the economic activities concentrated there; in rather more cases they did so at a time when they started to gain a competitive advantage." Similar statements note the hampering of industrial growth in southern Italy after unification removed tariff barriers. How provision of transport routes and establishment of commodity rates alone accentuate and perpetuate initial areal differences is set forth in my "The Role of Transportation and the Bases for Interaction," in *Man's Role in Changing the Face of the Earth*, ed. William L. Thomas, p. 865.

concentration in many countries[25] as in the American industrial belt, rather than building up to one point, as Haig theoretically postulated some years ago.[26] Reasons are obscure, but are probably related to resource distribution and processing requirements, as Haig implied, and to strategic values of different sites for different purposes. In practice, of course, the concentration is at points—cities—in an area.

The American industrial belt might also be considered as divided by the relatively less developed Appalachian barrier into two halves: seaboard, with New York as capital, and central, with Chicago as largest city. Added to this would be a third, smaller node, southern California, protected by distance from the industrial belt competition, with Los Angeles as center of gravity. This arrangement superficially appears to coincide with a rank-size rule—New York, Chicago, Los Angeles (especially if one considers Philadelphia an abberation!). However, the fit is too inexact, there is leakage in the system, especially through the port of New York and other ports, and, most important, there are national functions in the two halves of the industrial belt. New York and Washington, however, are of a higher order on the hierarchy for many national services involving intangible flows, including finance and government; conversely, the western half is higher for many large scale industrial and distribution activities involving tangible flows. Part of the mischief is caused by the center, New York, being on the edge of the country and thus not advantageously located for national assembly and distribution of many products.[27]

Prospects for the corner or fringe areas to compete with the core area are bleak, as many have noted, because of transport costs to the market and lack of local economies of scale, together with the assembly advantages of the core area. The marked concentration of innovation and decision-making in the core area also favors that area, insofar as an effect can be deduced.

Prospects for the fringe areas are not all hopeless. Room exists for use of ingenuity in discovering and promoting still more distinctive lures in the fringe regions; later generations of entrepreneurs in the core area may become complacent, or old plants less efficient as a result of inherited obsolescence. Even mere emptiness of a remote area can become a lure, as witness the establishment of the atomic bomb range in southern Nevada, the best site in the country for this purpose.[28] However, such advantages tend to be exceptions rather than the rule in the world today, in spite of exhaustion of many natural resources near the present centers.

25. Except for the primate city in some countries and former Colonial areas.

26. R. M. Haig, "Toward an Understanding of the Metropolis: Some Speculations Regarding the Economic Base of Urban Concentration," *Quarterly Journal of Economics* 40 (1926):179–208.

27. C. D. Harris, "The Market as a Factor in the Localization of Industry."

28. Within urban areas emptiness may provoke spectacular growth in furnishing large blocks of vacant land for large scale subdivisions, shopping centers, or even airports located on swampy ground.

A final factor of great importance is direct government intervention to even the competition, whether it be on the international level with capital and technical assistance, or particularly on the national level as the Cassa per Mezzogiorno—Fund for the South—in Italy. In the United States the same occurs with reclamation projects, highway allocations, and the like. Fringe areas do have votes; in Italy they are numerous, since one poor vote is more or less equal to one rich vote; in the United States our built-in system of equal state representation in the Senate gives added representation and political power to the West and South. To a degree this compensates for the unequal distribution of economic power, although I have no intention of pursuing the point. In any event the "early" stages of development often are subsidized by someone for welfare reasons, whether it be children by parents or schools, or new metals such as titanium and magnesium, or new weapons such as airplanes by the government. Fringe regions are no exception, with low income fringe regions in particular even receiving an extra measure of assistance for humane reasons.

The policy problem is to determine how far fringe assistance should go in terms of benefiting the total national welfare and not harming other sectors elsewhere. Internal country assistance tends to be greater than international, in spite of the fact that mobility of population and other factors are greater internally than internationally. At the same time that the Negro moves North, the federal government builds super highways in the South and pressure mounts for increased federal aid to education. As a result of these twin forces of increased mobility and increased national subsidy, disparity of per capita development within countries, at the least, will be less than would be the case without the operation of these forces. If one postulates that these forces will be reasonably effective, then intranational disparity in concentration will be measured more in absolute terms of population density, industrial build-up and the like, and not so much in per capita income. On the world-wide scene such a degree of anticipation is not so warranted, because of lack of either extensive mobility or sufficient extra-regional and worldwide assistance. However, some developments work in this direction, as with the proposals for the European Common Market, which include not only free exchange of goods and capital, but also of persons, within the six nations. Thus, Europe appears to be moving toward the pattern and scale of the United States.

CHAPTER 6   Geographic Theory and Underdeveloped Areas

In this short article, Ullman applies many of the principles he had developed earlier to the pattern of underdevelopment throughout the world, particularly in the Third World (see also chap. 4). The article is taken from Essays on Geography and Economic Development, *ed. Norton Ginsburg, University of Chicago, Department of Geography Research Paper no. 62 (1960), pp. 26–32. It provides insights into Ullman's wide understanding of various areas of the world, many of which he had personally visited.*

It is well recognized that societies and settlements have rich and poor, or relatively developed and underdeveloped members. Equally axiomatic is the concentration of relative poverty or richness in specific areas. The study of underdeveloped areas thus should have a distinctly geographic flavor in contrast to the more traditional study of underdevelopment by classes of society, political groups, or sectors of an economy. It is further assumed that underdevelopment by area generally is more than merely a coincidental grouping of a greater number of poorly developed persons in certain areas. The precise reasons for the great disparity in areal development, however, remain imperfectly understood. This lack of understanding is awkward, because for various policy and humane reasons, there is a desire to improve underdeveloped areas in the world, just as there is to improve the lot of poorer classes within a society.

Some of the distinctly geographical or spatial concepts applying to underdevelopment will be suggested briefly. Surprisingly they have not been noted previously in a systematic way. These spatial considerations in turn relate to the reasons for and the means of alleviating underdevelopment. Specifically the role of two key geographical concepts will be explored: (1) areal differentiation and (2) spatial interaction.

AREAL DIFFERENTIATION OF UNDERDEVELOPMENT

Commonly, underdeveloped areas are thought of as aggregates—that is, of whole countries. Such generalized aggregates give only a crude approximation to reality and mask significant differences. These generalizations, however, have some validity, because national states, as we shall note later, themselves create a considerable leveling or homogeneous effect. Underdevelopment, nevertheless, is generally sharply concentrated within one part of a country, whether it be southern Italy vs. northern Italy or the southern United States vs. the northeastern United States. In fact, realistic appraisal reveals that even within the underdeveloped portions considerable differences generally prevail, as between Apulia and the rest of the Mezzogiorno or southern Italy, or between the surrounding Appalachian hill country and the

Great Valley in the eastern United States. Even finer break-downs reveal that one village may be poor and an adjacent one relatively well off. In many cases the explanation appears to be related to relatively better natural resources, particularly when fertile plains areas are contrasted with mountains. In many cases, however, this explanation does not suffice.

The fact that underdevelopment may be concentrated in relatively small areas reveals much greater contrasts and therefore poses a greater problem than is commonly supposed. Commonly used national averages mask critical differences. Even using national states or political subdivisions as units may smooth out differences by more than a random averaging, since most such units are set up on a "long lot" basis combining developed and underdeveloped area. One has merely to consider the states of the United States, such as Michigan or Wisconsin with their southern margins in the productive industrial belt and their northern edges in the sterile, glacial scoured rocks of the north country, or the Great Plains states, with their eastern nodes in the productive prairies and their western borders out in the semiarid Great Plains.[1] In national states the contrast is probably even greater; the poor adhere to the rich, or vice versa.

What are the reasons for these sharp contrasts? Natural resources, although important in many cases, are not enough to explain all of the contrasts, whether it be on a national or local level.[2] Nor are cultural differences alone enough. Change in technology and social conditions often has an effect, although considerable lag is to be expected. Some examples from my own experience in Sardinia illustrate the problem. One village, Santu Lussurgiu, in the uplands, illustrates the changing nature of resources. It used to be a relatively prosperous community; the inhabitants were known as the "gentry of the mountains." Their prosperity was based on raising horses and oxen of high quality which were in great demand in the plains. Now with the irrigation of the plains and the adoption of tractors and motor cars two changes occur: the plains benefit more from the change in technology at the same time that the mountains lose their market for horses and oxen. The plains thus are becoming the relatively wealthy zones. All over Sardinia and Italy and in other parts of the world the removal of the twin scourges of malaria and banditry also has aided the plains at the expense of hilltop towns.[3] Still another Sardinian village, Cuglieri, is widely recognized as relatively wealthy; the wealth is popularly ascribed to two causes: (1) the surrounding olive groves on the rough hillsides which represent very real inherited capital,

"Transportation profoundly affects and in turn is affected by the distribution of population. Corridors of dense population characteristically cluster along many well established transport routes. However, transport is not an active agent and succeeds only if there is traffic potential. Because modern transport is so lacking in most parts of the world, all improvements tend to be considered good, but many are not of first priority. Examples are: construction of a railway as a matter of national pride when traffic potential does not justify it, building of many ports when consolidation of services in a few well equipped ports served by local highways would be more efficient, or building roads to open up new, unsettled territory when equally productive, already settled territory is without access to the outside world by modern transport or has unmaintained facilities." [Unpublished paper, "Transportation and Population Distribution," circa 1959]

1. Edward L. Ullman, "Regional Development and the Geography of Concentration," *Papers and Proceedings of the Regional Science Association* 4 (1958):184–85.
2. Norton S. Ginsburg, "Natural Resources and Economic Development," *Annals of the Association of American Geographers* 47 (1957):197–212.
3. Edward L. Ullman, "Sardinia—A Project for Economic Rehabilitation," *News Bulletin, Institute of International Education*, March 1958, pp. 36–42.

"The fact that people early sensed the importance of circulation and subsequently have done very little with it may prove something about the fundamental nature of circulation and the fact that we could not do much with it until techniques or thoughts were further developed." [Letter to Robert Mayfield, 29 July 1954]

4. Ullman, "Regional Development and the Geography of Concentration." For a similar concept note Gunnar Myrdal, *Rich Lands and Poor* (New York: Harper, 1957), pp. 26–27, which reads in part: ". . . play of the forces in the market normally tends to increase, rather than to decrease, the inequalities between regions," and "occasionally these favored localities and regions offer particularly good natural conditions for the economic activities concentrated there; in rather more cases they did so at a time when they started to gain a competitive advantage." How provision of transport routes and establishment of commodity rates alone accentuate and perpetuate initial areal differences is set forth in E. L. Ullman, "The Role of Transportation and the Bases for Interaction" in *Man's Role in Changing the Face of the Earth*, ed. William L. Thomas (Chicago: University of Chicago Press, 1956), p. 865.

and (2) the intelligence and energy of the inhabitants. Such contrasts as these are common in many underdeveloped countries and should be well worth recognizing and studying.

In any case concentration of development within countries is the rule even on a regional basis and may even be increasing, in spite of governmental policies to the contrary. The momentum of an early start is often a compelling circumstance, especially if it results in large scale market and development. This fact may signal the operation of a general localization principle in man's use of the earth: *initial location advantages at a critical stage of change become magnified in the course of development.*[4] Geographical differentiation starts out as a matter of homeopathic doses of mild concentration and winds up as a system of massive localization based on a wide range of internal and external economies of scale and cultural attributes. This direction continues until some radical new change is precipitated, and indeed may persist, even though new competitors arise elsewhere. The really significant geography thus becomes the concentration of man himself, his works, his institutions, and his inventive momentum.

This characterization appears to describe the greatest development in the two highly developed areas of the world, northeastern United States and northwestern Europe, which got the jump on other areas at a critical stage in the industrial revolution. The later emergence of Japan and Russia in part, however, requires other interpretations.

INTERACTION AND AREA DEVELOPMENT

The fact that development is unevenly spread in a country and in a manner not solely related to the natural endowment raises the question as to why development has not spread more evenly throughout a country. The problem is not analogous to a city where wealthy people or people of one class congregate by choice, although in some regional cases this is true, as in the settling of certain areas by people of one nationality. However, in a city the means of livelihood generally are not provided by the neighborhood of settlement, as in the countryside. A significant part of the explanation is related to the economies and momentum of concentration as noted previously. Some practices of political units also favor the equal spread of opportunity; this is one reason why national states have considerable validity as units of measurement. Common schools, roads, armies, markets, services of all kinds, relative freedom to migrate, and subsidy to the poorer regions, are all features of political area and work powerfully to even out the differences, although many remain and are even accentuated as Myrdal notes, in spite of this political uniformity.

A fundamental question to answer, therefore, is the relative "stickiness" of society, the resistance of certain areas to spread of innovations and improvements. Fundamental work on this question has been done by Torsten Hägerstrand, Edgar Kant, and others at the Lund School of Geography in Sweden. Hägerstrand has carefully plotted the spread of certain innovations in part of southern Sweden showing how they spread rapidly in certain areas. Independently, by means of plotting telephone connections and other measures, he relates this rapid spread to the greater degree of intercommunication in these areas. He is also able to predict independently by the use of Monte Carlo models and other methods using his general data, the actual spread of the innovation.[5] In the future it appears he may be able to do this merely by plotting the road network. This is a fundamental break-through and has strong implications for public policy and for the problem of spreading innovation and development in underdeveloped areas. Where, how far apart, and what type of demonstration projects should be established, for example?

Considerations of interaction and spread of innovation lead us to speculate on the role of typical introductions into underdeveloped areas—the Primate City, the plantation, the mine, or the Chinese and Indian merchant in Southeast Asia, for example. What effect have they had on development outside their immediate locale? Has their influence on the whole area been great or not? In all these cases the impression remains that these introductions have, until recently, remained largely isolated from the rest of the country.

Boeke, the Dutch economist, Furnivall, an English colonial scholar, and others have noted the separation between the Western and native worlds in colonies and speak, therefore, of "dual" or "plural" societies.[6] Boeke notes further that in Indonesia even before Western penetration cultural development was so essentially localized that the communities could be termed "small adjacent but non-communicating vessels." Nevertheless, "under the rule of the Javanese kings economic inner bonds fostered by home trade and internal migration were stronger" than under the later export, colonial orientation.[7] To take but one other example, local handicraft trade in specialized villages in the Philippines and Sardinia declines under the impact of competition with foreign or domestic machine-made products.

"Transportation is part of the complete system of interaction and circulation including transmission and communication with substitutability between modes, as telephone calls instead of trips.

All transportation and interchange are critical, interrelated, and will be investigated as part of the systems approach. The area with least capacity in relation to demand appears to be the intrametropolitan movement of people. In part this reflects the rapid and continuing growth of metropolitan centers and the shift in our economy from primary and secondary (agriculture and manufacturing) to tertiary (services) emphasis. Elsewhere in the system capacity generally appears to meet needs better, and in many segments, as on railroad line haul or rural highways there is over capacity." [Unpublished paper, "Transportation as Part of a Circulation System," December 1964.]

5. Torsten Hägerstrand, "Innovationsforloppet ur Korologisk Synpunkt," *Meddelenden from Lunds geografiska Institution*, Avhandlingar 25 (Lund, 1953). (The author is indebted to Professor Hägerstrand personally explaining the study to him. A statement in English is in preparation.)

6. J. H. Boeke, *Economics and Economic Policy of Dual Societies as Exemplified by Indonesia* (New York: Institute of Pacific Relations, 1953); J. S. Furnivall, *Colonial Policy and Practice* (Cambridge: Cambridge University Press, 1948; New York: 1956). See also *The Pattern of Asia*, ed. Norton S. Ginsburg (Englewood Cliffs, N.J.: Prentice-Hall, 1958), pp. 36–43; and Philip Wagner's discussion of the plural economy in chap. 3 of this volume.

7. Boeke, *Economics and Economic Policy of Dual Societies*, pp. 107–8.

"The problem of development or agricultural advancement of underdeveloped or backward areas is knotty. By and large I think most geographers agree that most of the poor areas of the world, and particularly in the tropics, provide the most expensive places in which to increase agricultural production. There is probably more pay dirt available for expansion or production in the corn belt than there is in Amazonia or tropical Africa. Basically tropical soils are unfertile; they are leached out. Exceptions are provided by some volcanic soils as in Java or from alluvial soils in the deltas of large streams. It is precisely in these spots in the tropics that large populations occur. Expansion elsewhere will probably require expensive adding of fertilizers or other ingredients to the soil in order to take advantage of the abundant heat and moisture. However, no simple statement can be made, especially by me since I am not an expert. These problems and opinions change with time as well as depending on the individuals who express them." [Letter to Carey McWilliams, 16 February 1954]

"Cities of India seem to be islands floating in a sea of rural poverty." [Letter to Brian J. L. Berry, 15 December 1967]

Western influences have mixed effects. The Primate City especially, even though its material and cultural standards differ drastically from the rest of the country, nevertheless has a great effect. Migrants swarm to it from all over the country. The principal newspapers and other media and institutions of all kinds center there. In some cases, as in India, one has an impression of two worlds—(1) the great cities with education, libraries, utilities, etc., connected with each other by strategic transport, but floating like islands in (2) a vast sea of rural villages without schools and facilities of any kind. Boeke indicates that Western influence even has a negative effect in one sense since it tends to divert the attention of the leading classes from their own society. The masses "unable to follow their leaders on their western way, thus lose the dynamic, developing element in their culture. Eastern culture in this way comes to a standstill and stagnation means decline."[8]

Similarly, plantations seem to have little effect. In Honduras, for example, one report notes that the economy remains predominantly what it was in 1821. It remains a land which has made relatively little economic progress in spite of a large-scale efficient banana export industry grafted onto the economy by United States capital early in the twentieth century.[9] A mine or rubber plantation in the Outer Provinces of Indonesia, it is argued, "carries on without touching native life at any point." Capital and labor "are both imported, the land was waste land, the product is all exported, and even the necessaries of life for the workers have to be brought from elsewhere. The whole concern is detached from its surroundings, although its indirect influence on the surroundings is penetrating."[10]

This same statement could be repeated all around the world, although, as noted, the Primate City, the plantation, the mine, and the Chinese, Indian, Levantine, Arab, or other foreign merchant all have had some effects. Labor, for example, migrates to the first three and then some flows back to the country with new desires. Servant girls, from Sweden to Sardinia, characteristically go from poor rural districts to work in the capitals and then return to their home districts to find husbands whom they attempt to make toe the new mark they have come to recognize. Chinese, Indian, and other blood ties appear in Southeast Asia in spite of apparent or asserted cultural isolation, and have an indirect effect on political as well as social and economic relationships.

The major reason why these introductions float without much effect, however, appears to be that they are too different. One might speculate that if society is composed of

8. Ibid., p. 39.
9. Vincent Checchi and Associates, *Honduras: A Problem in Economic Development* (New York: Twentieth Century Fund, 1959).
10. Boeke, *Economics and Economic Policy of Dual Societies*, p. 103.

great extremes—a few very rich and many poor, for example—there is likely to be less transmission of ideas and techniques than in a more democratic continuum. Hägerstrand has evidence that this appears to explain some slowness in innovation diffusion in parts of Sweden. Knowing the spatial aspects of the social and economic distribution of population thus would enable one to predict better the likelihood of diffusion.

A second vital consideration in underdeveloped areas is the provision of transportation, especially roads.[11] Railroads and sea transport provided a strategic net connecting major centers with each other or the outside world, but this network of connections has tended to remain separate from the mass of the internal economy and thus resembles the Primate City in effect. Furthermore, the provision of this transport tended to accentuate contrasts and made underdeveloped portions appear relatively, and in some cases according to some authorities absolutely, worse off. Moreover, in Sardinia, for example, it is apparent that the road net also was established first as a strategic net connecting cities and villages. In the process some much more heavily traveled routes linking villages with their immediately productive hinterlands were never improved sufficiently for wheeled vehicles, whereas another road through less productive territory and with less traffic, but running to another center, was improved.

Roads which permit their linkage with the rest of the country by means of wheeled transport generally are urgently sought by local villages in still poorer countries. They know the penalties of isolation. In any case, innovations, as well as people and products, are transmitted easier if movement is easier. This does not mean that transport automatically develops. It is a passive force, a necessary, not sufficient condition, but one with profound effects on spatial organizaton and underdevelopment.

CONCLUSION

Recognition of the important role of concentration and differentiation, or interaction and circulation, not only provides the beginnings of a theoretical geographical treatment of underdevelopment, but also contributes new insights to the whole problem of underdeveloped areas. Geographers in a sense have had these conceptual approaches implicit in their backyard all the time. Their explicit use and extension should greatly strengthen the geographic treatment of the problem.

The two concepts are related to the well-known geographic

11. For a general exposition of geography as spatial interaction, see Ullman, "The Role of Transportation and the Bases for Interaction," in *Man's Role in Changing the Face of the Earth*, ed. William L. Thomas.

terms of site and situation. In attempting to provide an explanation for the age-old puzzle of the growth of particular civilizations in particular places, for example, Toynbee in his "challenge and response" theory uses a site concept with a new twist—the challenging effect of a relatively poor environment.[12] Gourou, in reviewing this concept, poses the following query: Does the substitution of the effects of an unfavorable environment for the effects of a favorable one represent progress over previous interpretations based on environmental determinism?[13] He poses as an alternate possibility a situation concept, the rise of civilization in favored corridors for interaction, so that contact with other civilizations and contrasting ideas was facilitated, as in parts of Europe.

Why some places are developed and some are not still remains a mystery in many respects. We can, however, better isolate the cases for study by noting significant differences among them and can investigate by interaction techniques their relations with other places in order to understand their previous evolution and to aid policy determination for the future.

12. This is not entirely new. Witness the remarks of Herodotus: "Soft countries breed soft men"; or Montesquieu's: "the barrenness of the earth renders men industrious, sober, inured to hardship, courageous and fit for war," as quoted by David Lowenthal in his *George Perkins Marsh* (New York: Columbia University Press, 1958), p. 60. Marsh, as Lowenthal notes (pp. 60–64), invoked the same line of reasoning in some of his early writing to explain New England's virtues.

13. Pierre Gourou, "Civilisations et malchance géographique," *Annales, économies, sociétés, civilisations* (October–December 1949), pp. 445–50.

"Another way of stating this is that I am a spatial bookkeeper. I'm interested in the consequences [of theory and concepts] in terms of space, just as the economist is in terms of income or the demographer in terms of population, although none of us are 'pure'." [Letter to David A. Revzan, 1 June 1955]

"I have been guilty of blind empiricism myself at times. Blind empiricism sometimes is superior to cock-eyed conceptualism." [Letter to Edward A. Ackerman, 7 July 1953]

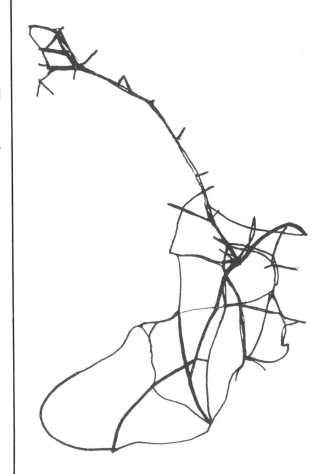

CHAPTER 7   Rivers as Regional Bonds: The Columbia-Snake Example

This article was published in the Geographical Review *in April 1951, pp. 210–25, and is a classic statement on the blending of "site" and "situation." The thesis is that rivers provide a bond—the essential glue of connection—between very diverse regions that require interaction. Ullman was particularly interested in streams crossing through mountains, which he named dioric streams (from the Greek* dia oros, *"through mountains"). It was the character of the crossing that was significant inasmuch as this set up the possibilities for interaction and use. The Columbia River, as it traverses the Cascade Mountains, is the prototype for all mountain-crossing streams.*

    The idea of situation was a persistent theme with Ullman; he simply applied it to various regions, in this case to the Pacific Northwest. As he was an instructor of geography at Washington State College in Pullman, Washington, from 1935 to 1937, the deserts and mountains of the Northwest may have triggered the concept of spatial interaction that he developed later.

1. W. M. Davis, "The Physical Geography of the Lands," *Popular Science Monthly* 57 (1900):157–70, reference on p. 158; reprinted in his *Geographical Essays,* ed. D. W. Johnson (Boston, New York: Ginn and Co., 1909), pp. 70–86, reference on p. 71.
2. The late Kirk Bryan, professor of physiography at Harvard, was the successor to William Morris Davis as geomorphologist with a genuine interest in geography. Professor Bryan read this manuscript, and although it was not the type of study he himself would have prepared, he

In 1900, William Morris Davis warmly approved a statement made by Guyot half a century earlier that it was not enough "to describe, without rising to the causes, or descending to the consequences."[*][1] The causes of rivers have been studied in detail by geomorphologists and others, but all too often their consequences have been left to the inadequate treatment of novelists and historians. This paper represents an attempt to rescue rivers from the geomorphologists on the one hand and the romancers on the other.[2]

Rivers have a definite relation to man, though it varies from place to place and from time to time. Nevertheless, their consequences, like their causes, are susceptible of systematic analysis. Rivers are more than conduits for water of a certain length, width, depth, and speed, for consumption, irrigation, power, or navigation. Along most of their courses they are hindrances to local cross-country movements, and sometimes they form effective international barriers.

More generally, however, rivers serve as regional bonds, because roads and railroads parallel their easy gradients and population is concentrated along their level or fertile valleys.

Two spectacular types of rivers exemplify this role of regional connection: rivers flowing through mountain ranges and rivers flowing across deserts. The Potomac is an example of the former, the Nile the classic example of the latter. The Potomac in places provides a water-level route for the two- and three-track main line of the Baltimore and Ohio Railroad on one bank and the coal-carrying Western Maryland Railway on the other, as well as for a canal (now no longer used) and a highway. The Nile is even more important as a corridor, not so much because boats use it and road and railroad follow it as because it supports a continuous band of population forming a living bridge—an unbroken series of steppingstones—across the North African desert. Numerous other examples of rivers crossing mountains or deserts can be cited: the several streams across parts of the eastern Appalachian Mountains; the Danube through the Iron Gate; the Yangtze through the gorge above Ichang; the Indus and the Tigris-Euphrates across deserts; and the Columbia-Snake

*Thanks are due the Milton Fund of Harvard University and the Office of Naval Research for financial aid in the preparation of this study.

across both mountains and deserts. Such streams provide routes across either or both of the two types of barriers defined by Hartshorne as "kinetic" when mountains and "static" when deserts.[3]

These rivers have a special character and a special function; they deserve special names. Rivers such as the Nile have been called exotic.[4] Such rivers are strange or foreign in that they decrease in volume downstream, because of evaporation and lack of tributaries. It would seem appropriate to extend this term to all streams crossing deserts, whether or not they lose volume, and thus avoid the need for a new term.

For streams flowing across mountains no satisfactory names exist. Generally such streams are given geomorphological names based on a hypothetical genesis. Not only is this practice unsatisfactory for our purposes, but interpretation of the origin often changes. For example, Powell in 1875 termed the Green River "antecedent" in its course across the Uinta Mountains.[5] Today it seems that many streams crossing mountains, including the Green, and also the Columbia and the Snake, are not "antecedent," but have some other origin (for example, "superimposition" or "pirate capture"). A more closely descriptive name is needed.[6] I have been unable to manufacture a term both self-explanatory and short derived from either English or Latin. I propose therefore that streams crossing mountains be called dioric, from the Greek dia oros, "through mountains."[7]

THE COLUMBIA-SNAKE RIVER SYSTEM

The Columbia-Snake river system probably crosses more markedly defined mountains and deserts combined than any other river system in the world. Two mountain ranges and two deserts are crossed within the United States; thus there are two dioric and two exotic stretches, and also some minor transitional reaches (fig. 7.1).

The lower Columbia, crossing the Cascades and the Coast Ranges, is an all-important transport route and regional bond, surpassing even the Potomac. The deep-cut Columbia Gorge provides the only water-level route through the entire Cascade Range (fig 7.2). Its easy gradients are followed by two transcontinental railroads and two highways. Its low elevations and openness to the mild West Coast give it a warmer winter climate than that of the snow-covered mountain passes, a vital factor in keeping its transport lines open in winter. It also serves (1) as a passageway for weather from the East—because of pressure differences, easterly gales some-

was kind enough to express his appreciation and to make suggestions for more detailed studies. In a small way, therefore, this paper is a response to the intellectual challenge given me by two geomorphologists, Bryan and Davis, but more directly to the stimulation and kindly interest of Professor Bryan.

3. Richard Hartshorne, "Suggestions on the Terminology of Political Boundaries," Mitt. Vereins der Geographen an der Univ. Leipzig 14/15 (1936):180–92. See also the abstract in the Annals of the Association of American Geographers 26 (1936):56–57.

4. P. E. James, An Outline of Geography (Boston, New York, etc., 1935), p. 29.

5. J. W. Powell, Exploration of the Colorado River of the West and Its Tributaries (Washington, D.C.: Smithsonian Institution, 1875), p. 163. See also Davis, "The Physical Geography of the Lands," p. 166 (Geographical Essays, p. 81); and his "The Rivers of Northern New Jersey, with Notes on the Classification of Rivers in General," National Geographic Magazine 2 (1890):81–110, reference on p. 82 (Geographical Essays, pp. 485–513, reference on p. 486).

6. Cf. Henri Baulig, "Questions de terminologie, I," Journal of Geomorphology 1 (1938):224–29, reference on p. 225.

7. This term was coined for me by my father, B. L. Ullman, professor of classics, University of North Carolina. I hesitate to add new terms and thus remove geography from the understanding of intelligent laymen, especially since at least sixty different types of streams have already been named (see E. H. Webster, "Types of Rivers and Mountains," Journal of Geography 39 (1940):206–07. Many of the names, however, seem less significant than dioric.

I have (perhaps mistakenly) resisted the temptation to coin a logical parallel to dioric to replace exotic. Dixeric is one possibility, but it is not as suitable as dieremic, which means "through deserted places" and was applied to the Sahara by Herodotus. Unfortunately, both terms present problems for English pronunciation, particularly the latter. Complete logic would use diaphragmatic,

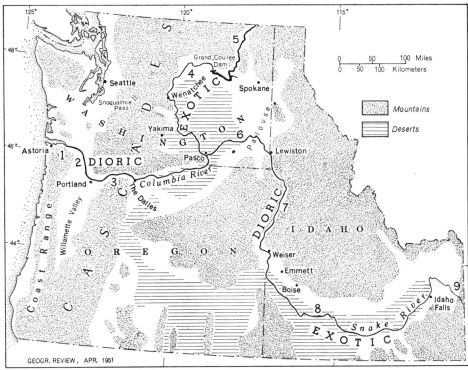

Fig. 7.1. The Columbia and Snake Rivers in relation to mountains and deserts, showing *dioric* and *exotic* stretches. Principal subdivisions are either transition or core stretches: (1) the dioric Columbia crossing the Coast Ranges; (2) the Columbia crossing the Puget Sound-Willamette Valley trough; (3) the main dioric Columbia through the Cascades; (4) the exotic Columbia skirting the mountains and crossing the desert of central Washington; (5) the Columbia up to (and beyond) the Canadian border running through mountain and foothill valleys; (6) the Snake from the Columbia to Lewiston, which might be considered a combination dioric-exotic stream, since it crosses the desert near Pasco and is also intrenched in a gorge beginning about 25 miles above the confluence; (7) the dioric Snake above Lewiston through gorges culminating in Hells Canyon; (8) the exotic Snake to a point above Idaho Falls; (9) the canyon of the upper Snake leading into Wyoming.

"through barriers," as a generic term for all streams crossing barriers. There is also much merit in still another term, *transverse*, formerly used by geologists and others for streams crossing the grain of the country, though not for streams crossing deserts, so far as I know.

8. D. C. Cameron, "Easterly Gales in the Columbia River Gorge during the Winter of 1930–1931," *Monthly Weather Review* 59 (1931):411–13.

times blow down the gorge in winter and reduce ground fog at Portland to shorter duration than at Salem or Seattle;[8] (2) as a flyway for birds and airplanes; and (3) as a watercourse for vessels drawing twenty-six feet as far as The Dalles, 188 miles from the mouth, and for barges drawing five to seven feet as far as the mouth of the Yakima River, 139 miles above The Dalles at Pasco. Above Portland, however, little traffic is carried except inbound petroleum products. Below Portland the channel is thirty-five feet deep and is a much-used water route to the sea, though only one railroad and one highway parallel the river across the Coast Ranges.

In contrast, the upper Columbia, crossing the relatively level desert of central Washington, is of little importance as a bond, because its water has not yet been used for irrigation. At present, only tributaries of the Columbia, such as the Yakima and the Wenatchee, are used significantly for irrigation, creating populous valleys extending out into the desert from the eastern flank of the Cascades. The Columbia itself is intrenched in a gorge a few hundred feet deep, which requires construction of large dams or expensive pumping before water can be diverted from it. As a result, central Washington is empty desert between the Cascades and the relatively populous wheat-farming belt of eastern Washington and northern Idaho, where precipitation is greater (fig. 7.3).

The effect of the Snake is just the reverse. So precipitous are the banks of its canyon that only a pack-animal trail has been built along this dioric section (fig. 7.4). The gorge is a barrier forcing transportation into the adjoining mountains. Hells Canyon on the Snake is more than seven thousand feet deep, the deepest canyon in the United States, exceeding even the Grand Canyon of the Colorado. It has not attracted attention because it can be reached only by trail or by a combination of an eighteen- to twenty-five-hour motorboat trip up the rapids from Lewiston to Johnson Bar and a trail trip for some twenty miles up from this head of so-called "navigation."[9] In the deeper parts of the canyon a few scattered ranches are based on the occasional alluvial fans debouching into the gorge. These are practically the only level areas with readily available water on which alfalfa or other fodder needed to carry stock through the winter can be raised; feed cannot be imported because of lack of transportation. The number of cattle is a direct function of the limited area of the alluvial fans, an almost inconceivable situation in a modern transportation corridor.

The fourth section, the exotic upper Snake, resembles the Nile in its effects: in contrast with the upper Columbia, it supports a chain of irrigation settlements along its banks (fig. 7.5). Although the upper Snake is slightly intrenched, its gorge is much shallower than the gorge of the Columbia across the desert of central Washington; consequently, dams have been built at some places, such as American Falls (fig. 7.6). Elsewhere, as at Boise, the water of tributaries is used, but the land irrigated is a part of the Snake River Plain.

Thus one dioric stretch of river, the lower Columbia, is an all-important bond, and the other dioric stretch, the Snake Canyon, is a barrier; conversely, one exotic stretch, the Snake

"I guess the major point that I was making in my studies is that a zone tends to develop. Nature and man appear to dislike an arbitrary boundary and spill over it to a certain extent, but the boundary definitely still slows things down. Incidentally, the Woonsocket area boundary—the northern boundary of Rhode Island—is not strictly speaking straight but has some wiggles in it developed not only from poor surveying but the poor survey was adjusted to by the settlers before a good survey came along and so the poor survey was perpetuated, as I recall.

I also wrote a historical geography of the eastern boundary of Rhode Island which appeared in the Research Studies of the State College of Washington and is referred to in my Geographical Review article. In that I related the boundary dissension to the fact that it did not take in all the Narragansett Bay Watershed. I suppose I was a bit influenced by environmental determinism when I wrote that, just as the approach I took in the Geographical Review article was a landscape one influenced by that particular emphasis in American geography at that time. If I were to do the study today I suppose I would concentrate a little more on an interaction approach, and the overspill of metropolitan areas. In other words, I would tie it in with trade areas or commuting areas of the cities a bit more."
[Letter to William C. Metz, 29 November 1973]

9. For a general description of the canyon, see O. W. Freeman, "The Snake River Canyon," *Geographical Review* 28 (1938):597–608.

Fig. 7.2. The Columbia Gorge at Bonneville Dam, looking south to Mt. Hood. (Photograph by the Bonneville Power Administration.)

Fig. 7.3. Empty, arid country typical of much of southern Idaho and central Washington. Potentially irrigable sagebrush lands near Boise. (Photograph by B. D. Glaha, Bureau of Reclamation.)

Fig. 7.4. Irrigated land in Snake River Plain. View north toward Emmett, Idaho, from top of Freezeout Mountain. (Photograph by B. D. Glaha, Bureau of Reclamation.)

"Fred Richardson sent me a copy of the note he sent you commenting on the terms I used in 'Rivers as Regional Bonds' in the April Review. . . . He may well be right about a term such as 'dioric.' I honestly have no particular pride of authorship in that, and terms such as that should be sort of voted on by the profession." [Letter to Wilma Fairchild, 7 June 1951]

in southern Idaho, is a significant bond, and the other exotic stretch, the upper Columbia, is not now a bond.[10]

The role of these diverse sections can be better understood if the distribution of transportation and population over the whole region is considered.[11]

TRANSPORTATION AND POPULATION

Rail traffic is relatively heavy through the Columbia Gorge on two main lines and relatively heavy across the desert of Washington, completely absent through the Snake River Canyon and fairly heavy on the Oregon Short Line (Union Pacific) across the Snake River Plain (fig. 7.7). A map of main freight traffic flows would result in about the same relative picture as passenger movements in 1943, except on the main line of the Milwaukee across Washington, where passenger density was relatively lower than freight density.[12]

The flow across the Cascades to Puget Sound, however, is heavier than that through the Columbia Gorge. Initially, Puget Sound furnished a much better water route through the Coast Ranges than the Columbia River, and partly for this reason (as well as because of the availability of fairly low

10. In addition to this primary fourfold division, note the minor transitional subdivisions listed in the caption to figure 7.1.

11. For an excellent general discussion of the Columbia River as a regional bond and the general strength of east-west ties in the Pacific Northwest, see Charles McKinley and Blair Stewart, "Definition and Regionality of Pacific Northwest or Columbia Region," in *Regional Planning, Part I: Pacific Northwest* (Washington, D.C.: National Resources Committee, 1936), pp. 95–131 (section 3).

12. Data for freight flows from Copeland traffic-density charts. Facilities also mirror the traffic pattern (compare the map of "U.S. Railroads," figure 1 (pp. 244–45) in E. L. Ullman, "The Railroad Pattern of the United States," *Geographical Review* 39 (1949):242–56.

Fig. 7.5. American Falls storage dam (background) and Idaho Power Company dam (foreground) on the Snake River. (Photograph by F. B. Pomeroy, Bureau of Reclamation.)

Fig. 7.6. Snake River Canyon, Idaho-Oregon, about 6,000 feet deep at this point, close to Hells Canyon. (Photograph by Harry Clements.)

passes across the Cascades), Seattle, Tacoma, and other Puget Sound ports became more important than Portland and Astoria on the Columbia and have attracted three transcontinental lines. However, improvements in the channel of the Columbia, particularly across the bar at the mouth, have lessened the river's handicap in recent years and provide Portland with thirty-five feet of water, enough for the largest commercial vessel on the Pacific.

Highway traffic is fairly heavy through the Columbia Gorge and across the Snake River Plain and, conversely, completely absent through the Snake River Canyon and very light across the desert of central Washington (fig. 7.8). More traffic crosses the mountains from the Yakima and Wenatchee Valleys to the Pacific Coast than crosses the relatively level Columbia Basin toward Spokane. Highway traffic, in contrast with rail traffic, is short-haul; consequently, heavy highway densities correlate with dense population.

Within the basin, and east of the Puget Sound-Willamette trough, concentration of population is noteworthy along the Snake River Plain (fig. 7.9) and sparse on the Columbia Plateau. The Snake River Canyon has practically no population, but the Columbia Gorge has a considerable clustering. A map of irrigated areas (fig. 7.10) explains the concentration: the Snake River Plain has the second-largest amount of irrigated land in the United States, being exceeded in acreage only by the Central Valley of California. Other parts of the basin have scattered projects, particularly tributary valleys of the Columbia, principally the Yakima, with its associated dense population.

NEW DEVELOPMENTS IN TRANSPORTATION

The impact of the different sections of the river has varied. For example, the emigrant route to the Pacific Northwest in 1845, with its first rudimentary improvements, did not follow the water-level route through the Columbia Gorge, because of the Cascades rapids, but swung through a pass south of Mt. Hood.[13] As travel increased, however, the route was soon changed to follow the gorge, utilizing steamboats on the Columbia and a fourteen-mile portage railroad round the Celilo Falls completed in 1863.[14] Still later it was found feasible to build railroads and highways along the Columbia.

The breaching of the Cascades behind Seattle by three transcontinental railroads and three highways, one of them open all year, has been the major challenge to the relative importance of the dioric Columbia as a bond. Construction of

13. Federal Writers' Project, "The Oregon Trail" (American Guide Series, New York, 1939), p. 143.

14. Ibid., p. 138. It should be noted that the gorge of the Columbia had served as an Indian route for centuries (J. N. Barry, "The First Explorers of the Columbia and Snake Rivers," *Geographical Review* 22 (1932):443–56, reference on p. 448).

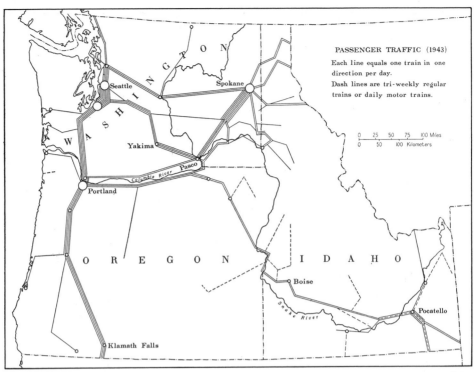

Fig. 7.7. Railroad passenger-traffic density. (Computed from "The Official Guide of the Railways," New York, 1943.)

Fig. 7.8. Highway traffic density. (Source: "Regional Planning, Part I, Pacific North-west," National Resources Committee, 1936.)

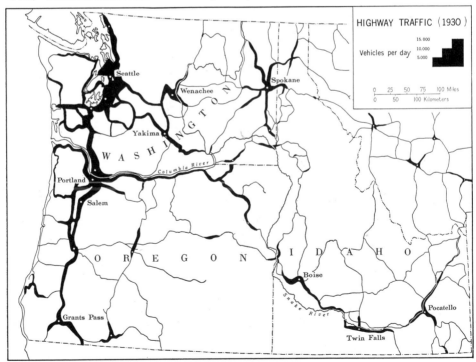

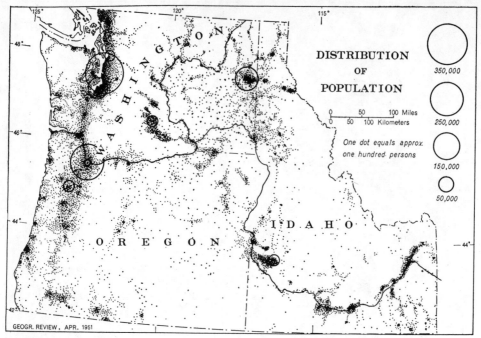

Fig. 7.9. Distribution of population. (Source: National Resources Planning Board map in "Atlas of Agriculture: Pacific Northwest Region," U.S. Dept. of Agriculture, Pacific Northwest Regional Committee for Post-War Programs, Portland, Oregon, n.d.)

Fig. 7.10. Present and proposed irrigated lands. (Source: "The Columbia River," Bureau of Reclamation, Washington, D.C., 1947.)

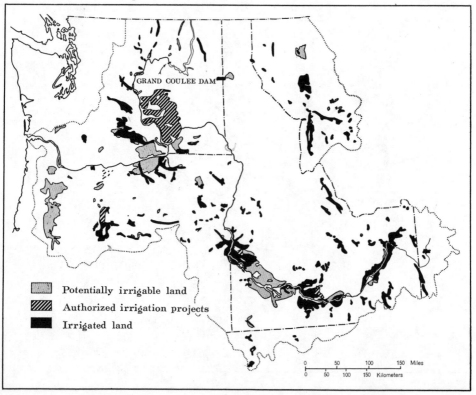

15. "The Cascade Low Level Tunnel" (Cascade Tunnel Report), Washington State Planning Council, Olympia, 1936, pp. 39–42; Ole Singstad, "Snoqualmie Tunnel Plan Provides Low-Level Route through Cascade Mountains," *Civil Engineering* 17 (1947):24–27 and 70. Improvements have been continuous. In 1949, for example, radio-operated barriers were installed to close the highway when snowslides occurred (K. G. Miller, "Barriers Close Snow Blocked Passes," *Better Roads* (1949):31–32, reference on p. 31).

16. Singstad, "Snoqualmie Tunnel Plan."

17. "Cascade Tunnel Project Remains a Live Issue," *Engineering News-Record* 140 (1948):21. See also *Engineering News-Record* 138 (1947): 972.

18. H. M. Chittenden, "A 30-Mile Railway Tunnel under the Cascade Mountains," *Engineering News* #76 (1916):1190. See also "First Report of the Cascade Tunnel Commission," State of Washington, 1926; reproduced in the "Cascade Tunnel Report", pp. 17–23.

19. *Railway Age*, vol. 127, no. 12, 17 September 1949, pp. 507–509. It should be noted that the Northern Pacific and Milwaukee tunnels, respectively at and near Stampede Pass, are much easier and needed no such outlay. Perhaps a cultural factor, the desire of Jim Hill to build a road competitive to the Northern Pacific and yet not run through the same local territory, may be the principal explanation for the choice of the costly northern route. Competition of privately built railroads, a distinctive feature of the United States, has left its stamp on the geography of the region just as surely as relief or climate has.

tunnels, electrifying or dieselizing of railroads, and development of improved snow-removal equipment for highways and railroads have reduced the barrier effect of the Cascades. The mountains were fortunately endowed by nature with low passes, such as Snoqualmie (3,010 feet), the main highway route, but deep snows (average annual fall at Snoqualmie, 411.6 inches) forced automobile traffic to detour in winter via the dioric Columbia until 1930, the first year Snoqualmie was kept open all winter. As early as 1934 the state had radio-equipped rotary plows and other snow-removal equipment, which kept the pass open except for a few short periods, at an annual cost of $40,000–$70,000. (During the severe winter of 1945–1946 snow removal cost about $100,000.) Grades had been reduced to a 5 percent maximum, and curvature to easy standards, so that driving conditions were equal to those on the highways along the Columbia.[15]

The Snoqualmie highway is still not completely satisfactory, as witness Governor Wallgren's proposal in 1947 for a 2.03-mile tunnel under Snoqualmie Pass below twenty-five hundred feet to avoid most of the snow and snowslides and reduce grades. The estimated cost was $21,940,000; the tunnel would be the longest vehicular tunnel in the world.[16] The proposal was indefinitely postponed by the 1947 Washington Legislature as unnecessary, but it may be raised again.[17] In the meantime a water-level highway along the south bank of the Columbia has been constructed to provide a straighter and easier-gradient route than the original scenic highway along the side of the gorge, so that the Columbia route has regained some of its relative advantage.

Several years before these recent improvements in rail and highway routes were made, a somewhat visionary thirty-mile low-level tunnel under the Cascades had been proposed, to enable Seattle to compete effectively with Portland and its water-level route to the interior.[18] A series of major disasters from landslides on the Great Northern focused public attention on the dangerous crossing of the Cascades. Completion of the Cascade Tunnel of the Great Northern in 1929 climaxed a $25,000,000 improvement program that largely eliminated the dangers on its line. As late as 1949, however, the Great Northern finished another short-line adjustment (one mile long), costing about $100,000,000, primarily to ease curves. Substantial conquest of the Cascades has been achieved, but only at considerable cost.[19]

At various times in the past construction of a railroad through the Snake River Canyon has been proposed. Initially the canyon was investigated as a water-level route for the Oregon Short Line (Union Pacific) from the East to Portland.

The distance, about double that of a direct route across the Blue Mountains, and the excessive curvature that would be necessary to follow the sinuosities of the canyon nullified its gradient advantage, in spite of the climb to forty-two hundred feet required for the line eventually built across the Blue Mountains. In addition, construction apparently would have been far more costly.

The canyon does not connect sufficiently important traffic sources to justify the cost of a railroad. Later, when the traffic potential had increased (but still not enough to justify a railroad) and Idaho wished to connect its northern and southern parts for political reasons, the automobile had been perfected, and the north-south highway was constructed through the mountains parallel to the gorge but some twenty to fifty miles away. A highway is much less sensitive to grades than a railroad and could be built through the mountains relatively inexpensively, whereas a railroad would have had to follow the costly easy-gradient route along the winding canyon bottom.

NEW DEVELOPMENTS IN IRRIGATION

The greatest changes in the use of the rivers are associated with the development of irrigation. Extensive irrigation on the Snake River Plain, for example, dates only from 1880–1900. Before that time most of the plain was empty.

Irrigation projects (see fig. 7.10) indicate still more changes in the future, the most important of which is the reclamation of about a million acres of the Columbia Basin by use of water from Grand Coulee reservoir (Lake Roosevelt). Grand Coulee Dam is the largest concrete dam in the world, and its construction had to wait until technology and capital were adequate to erect a dam 550 feet high above bedrock, 500 feet thick at the base, and 4,300 feet long. Even with this height and additional average pumping lift of 290 feet will be required. The sustained high flow of the Columbia throughout the summer, unique among major rivers of the western United States, ensures both adequate irrigation water and production of low-cost secondary power for pumping during the growing season.[20] The summer peak of the Columbia, reflecting its origin in distant glacial snows, considerably reduces the handicap resulting from its intrenchment.

The effect of the Columbia Basin irrigation project will be to close most of the population gap across the desert part of the northern Columbia Plateau from the Cascades east to the vicinity of the fertile Palouse of easternmost Washington; the

20. "The Columbia River: Comprehensive Report . . . ," February 1947, Bureau of Reclamation, U.S. Department of the Interior (1948), pp. 215 and 234; also O. W. Freeman, "Development of the Columbia Basin Reclamation Project," *Yearbook, Association of Pacific Coast Geographers* 9 (1947):15–22.

Columbia will then have created a band of population comparable to that of the Nile in effect, however, different in shape and engineering development.

On a smaller scale (in acreage irrigated), dams are proposed at places along the main stem of the Snake, such as the Bruneau Project (southeast of Boise), with the large Bliss Dam, 500 feet high and 4,230 feet long, where the river is intrenched several hundred feet deep.[21] The effect will be to close some of the gaps in the settlement pattern across southern Idaho and make still stronger the chain of interconnection (fig. 7.10).

EFFECT OF NEW DEVELOPMENTS OF THE AREA-CONNECTING ROLE

Thus the two dioric sections seem to be gaining little if any relative importance as bonds while the two exotic sections are becoming more important. The dioric Columbia is being made navigable by the construction of dams, but whether this will enable it to compete with the improvements in overland travel is questionable. The dioric Snake (the canyon) seems to be completely left out of any future strengthening of ties, in spite of the possibility that dams will be constructed here soon, even on a grand scale, though the absence of transportation makes the construction of large dams costly[22] (but, conversely, the absence of a railroad in the bottom of the canyon permits construction of a high dam that will flood the gorge).[23]

The all-important role of electricity, the chief product of the enormous dams built and projected, has been purposely omitted from this discussion, even though the flow of electricity is the Pacific Northwest's equivalent of the heavy coal movements that are the backbone of eastern America's traffic. It has been omitted because an essentially uniform, "postage stamp" rate regardless of distance was adopted for the Pacific Northwest grid, so that the region would benefit uniformly and industrial development would not be centralized at the damsites; the scenic attractions of the Columbia Gorge would thus be preserved and flexible planning in the region be made possible. Nevertheless, the net effect has paradoxically been to favor centralization. Centers already developed, particularly the cities of Portland, Seattle, and Spokane, have grown even larger because of agglomeration benefits.[24]

The government's huge Hanford plutonium works may reflect other considerations. Hanford consumes large amounts of electricity, and it is therefore strategically located on the

21. "The Columbia River," p. 153.

22. W. G. Hoyt, "Water Utilization in the Snake River Basin," *U.S. Geological Survey Water-Supply Paper 657* (1935), p. 247.

23. This rugged area is so spectacular that its very barrier quality may result in its being developed as a national park and thus acting as a magnet for tourists.

24. The interesting consequences of the Columbia in Canada have also been purposely ignored. The Canadian border is close to the end of the exotic Columbia and of the lake behind Grand Coulee Dam. The Columbia in Canada crosses no large deserts and runs parallel to the grain of the country over most of its course. Other changes in water use, such as the effect of dams on fish runs, also are not treated here, because of their minor relevance.

Columbia (near Pasco) between Bonneville and Grand Coulee, but closer to the latter's greater capacity. The Hanford works also require vast amounts of cooling water, so that location on a large stream is necessary. The Columbia, with its cold water and enormous minimum flow regulated by Grand Coulee Dam, is not deeply intrenched at this point, so that water can be easily taken. Isolation in the desert is also desirable (the works reservation alone covers 631 square miles).[25] The net result is another strengthening of the connecting role of the exotic Columbia.

SOME GENERAL COMMENTS ON DIORIC AND EXOTIC STREAMS

1. Recognition of the dioric and exotic character of the Columbia-Snake rivers provides an orderly and meaningful basis for describing this important river system, a method never before used in the voluminous literature on the basin. Other river systems may have entirely different characteristics, which should form the basis for a systematic description, in contrast with the generalized or genetic methods now in vogue in popular and most scientific literature.[26]

2. Dioric and exotic streams are so striking that they deserve special names so that their significance can be appreciated. The terms might be extended to cover other types of bonds across mountains and deserts; for example, more or less level passageways through mountains, such as the Mohawk Corridor break through the Appalachians (but not a mountain pass), or a series of relatively well-watered grazing lands or of oases based on underground water, such as the Fertile Crescent and its neighboring routes between Syria and Mesopotamia[27] or the old silk route following the discontinuous series of oases along the southern edge of the Tarim Basin in Chinese Turkestan.

3. The stage of technology, level of demand, and attitudes and objectives of the people determine the significance of dioric and exotic streams, just as they do for all other elements of the natural environment. One period's asset is another's liability and vice versa. Advances in technology making possible the construction of large dams will enable the exotic Columbia to fill an area-connecting role, whereas the shift from rail to auto has apparently doomed the dioric Snake to a barrier role. Recognition of dioric and exotic streams therefore implies no deterministic bias; it is merely a red flag calling attention to significantly greater possibilities than those inherent in normal mountain or desert crossings or in "normal" rivers elsewhere.

25. D. J. Brumley, "Atomic Energy Town of Richland, Wash., Grows from 250 to 25,000 Population," *Civil Engineering* 19 (1949):31–35, reference on p. 31.

26. William Morris Davis and his followers have constantly emphasized that no one will remember the name of a physiographic feature unless the name indicates the genesis. The argument is scarcely defensible on logical grounds, and certainly not for dioric and exotic streams.

27. Cf. M. Rostovtzeff, *Caravan Cities*, trans. D. and T. Talbot Rice (Oxford: Oxford University Press, 1932), pp. 2 and 95 and *passim*.

At the present stage of American history and technology it appears that dioric streams are losing some of their relative area-connecting role and that exotic streams are gaining in importance as regional bonds. The passing of the canal era in the 1840s meant less reliance on water-level routes such as the Potomac and the Mohawk. Some recent improvements in railroads, particularly the use of diesel locomotives, have enabled them to conquer grades more easily than before, but the greatly increased weight of trains has worked in the opposite direction. More important has been the rise of means of overland movement that are relatively independent of terrain—the automobile and the highway, the airplane, pipelines for oil and natural gas (which already cover more route miles than railroads), and, finally, high-tension electric lines. Furthermore, large-scale, multiple-purpose river and dam projects are now technically feasible, and politically popular, and thus exotic rivers are being considered for use all over the world as never before.

Trade Centers and Tributary Areas of the Philippines   CHAPTER 8

An integral, but generally ignored, feature of the geography of an area is the delimitation of nodal or functional regions — specifically, tributary areas or spheres of influence.[*][1] In a sense these are the real human regions. To establish a definitive set of such regions is difficult because of overlapping functions, progressive change, and the lack of precise data. For this reason even most well-developed countries have poor or no maps of this fundamental phenomenon. However, by judicious use of various indicators, a useful preliminary framework can be established.

PROCEDURE

Formidable difficulties hamper the delineation of a hierarchy of nodal centers and tributary areas, especially in underdeveloped areas. The Philippines are no exception. In the Philippines, as in most other underdeveloped areas, increased commercialization and the rapid building of roads are apparently creating a firmer set of organizational regions. But it is an enormous assignment, even with good data, for a single research worker with limited time to cover a country of large and scattered extent with twenty-two million people (fig. 8.1) and about two hundred major trade centers and thousands of agricultural villages.

Two key problems confront the researcher: (1) establishing objective criteria for classifying the centers; and (2) finding objective data for bounding the tributary areas. In the Philippines even census population figures are poor indicators of center size because population counts are made for *poblaciones*, political divisions that correspond poorly to the actual urban centers.[2] On the other hand, one new source of data is avail-

Ullman visited the Philippines as part of a Stanford University research project from March through September of 1956. In his capacity as Transportation Consultant on the project it was his mission to prepare a transportation survey of the Philippines for the National Economic Council. This paper is the professional result of that assignment, prepared for the geography community.

The paper presented here is not simply an inventory report of the condition of transportation in the Philippines, although that certainly is revealed, but rather an application of central place theory in a particular laboratory. In fact, the Philippines represent one of the most difficult areas in the world in which to apply the concept of central place. Most researchers have selected relatively homogeneous physical regions—such as Southern Germany in the case of Christaller, or Iowa in the case of Berry—while Ullman applied the concept to the island character and regional diversity of the Philippines.

*Field work for this study was done while the author was a member of a mission sent out by the Stanford Research Institute in 1956 to make a six-month transportation survey of the Philippines for the Philippine National Economic Council and the International Cooperation Administration. This resulted in a seven-volume report by R. O. Shreve, H. E. Robison, R. E. Arnold, J. W. Landregan, J. A. McCuniff, and E. L. Ullman, "An Economic Analysis of Philippine Domestic Transportation" (Stanford Research Institute, Menlo Park, Calif., 1957). Grateful acknowledgment is made to these agencies and to numerous individuals and officials in the Philippines who cooperated in the field investigations.

1. Cf. F. H. W. Green, "Community of Interest Areas: Notes on the Hierarchy of Central Places and Their Hinterlands," *Economic Geography* 34 (1958):210–26.

2. A partial exception is provided by Spencer's careful estimates of chartered-city populations for 1958, covering twenty-nine cities, most but not all of the larger cities. J. E. Spencer, "The Cities of the Philippines," *Journal of Geography* 57 (1958):288–94. Where appropriate, these are used here in the text and also in the separate table referred to in footnote 5.

Ullman considered this article to be one of his greatest achievements. He had used innovative and creative techniques to apply the theory of central place in his study, and the article provides excellent documentation of the validity of the concept. If, indeed, the theory could be applied in this extreme example, surely it must have universal applicability.

The article was published in the Geographical Review *in April 1960 (vol. 50, pp. 203–18).*

3. F. H. W. Green, "Urban Hinterlands in England and Wales: An Analysis of Bus Services," *Geography Journal* 116 (1950):64–88; "Community of Interest Areas in Western Europe—Some Geographical Aspects of Local Passenger Traffic," *Economic Geography* 29 (1953):283–98; and other papers by the same author. Sven Godlund, "Bus Service in Sweden," *Lund Studies in Geography*, series B, no. 17, 1956; and "The Function and Growth of Bus Traffic within the Sphere of Urban Influence," ibid., no. 18, 1956. (A much larger work, in Swedish, of which these two papers are summaries, was published in 1954.)

4. This method is similar to that employed for the TVA. H. V. Miller, "Effects of Reservoir Construction on Local Economic Units" [in the Knoxville-Chattanooga Area], *Economic Geography* 15 (1939):243–49.

able: maps on scales ranging from 1:200,000 to 1:400,000, prepared by the Bureau of Public Highways, of traffic flow on the roads of each of the fifty-two provinces for the years 1953 and 1954. Not only were these maps the principal basis for delimiting trade areas, but they were also a key to the recognition and classification of centers by the concentration of traffic flows and as indicators of the size of hinterlands. Since the main traffic on these roads is buses, the procedure in delimiting trade areas was similar to that followed for the well-known tributary-area maps based on bus traffic for England and Western Europe by F. H. W. Green and for Sweden by Sven Godlund.[3]

In addition to traffic flow, other consistent island-wide data used to classify the centers were branch plants and warehouses for the major soft-drink distributor (Coca-Cola is the one commodity distributed everywhere!); headquarters, main installations, and depots for the three oil companies distributing throughout the islands; provincial capitals; Chinese Chambers of Commerce (since the Chinese control most of the retail trade); and, as a poor indicator, population. A variety of specific data were also of minor use for individual localities—for example, port facilities, military installations, mines, sawmills, factories, and sugar centrals. For Manila several other unique indicators were used.

As preliminary field work for delineation of the trade areas, throughout the islands, merchants, presidents of Chinese Chambers of Commerce, and other businessmen in a major center were asked to delimit the trade area of their center for their specialty. Owners or managers of small stores in rural barrios at the edges of the trade areas were next interviewed and asked the same questions as a check.[4] The combined results were then compared with the Bureau of Public Highways traffic-flow maps, which were based on traffic counts. The correlation was in general remarkably close; the boundary of a trade area coincided with the traffic divide between centers. Because of the close correlation, the traffic-flow maps were also used for numerous places not visited personally (fig. 8.2).

The main base for the trade areas was thus provided by the traffic-flow maps. For some areas precise delimitation was difficult because the traffic counts were not sufficiently closely spaced. For some areas without extensive road networks, water traffic between ports furnished clues. Launch routes, stream patterns for native canoes, and trails were also used as guides. In still other cases, bus routes and other data provided additional information.

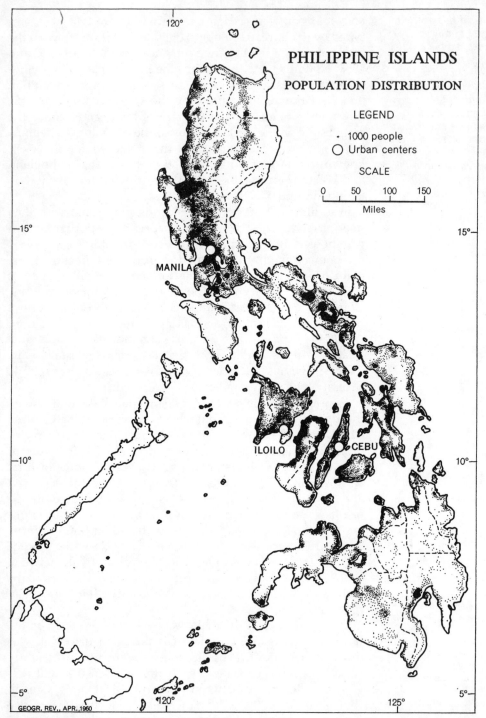

Fig. 8.1. Population distribution, 1950. (Source: Philippine Studies Program, University of Chicago.)

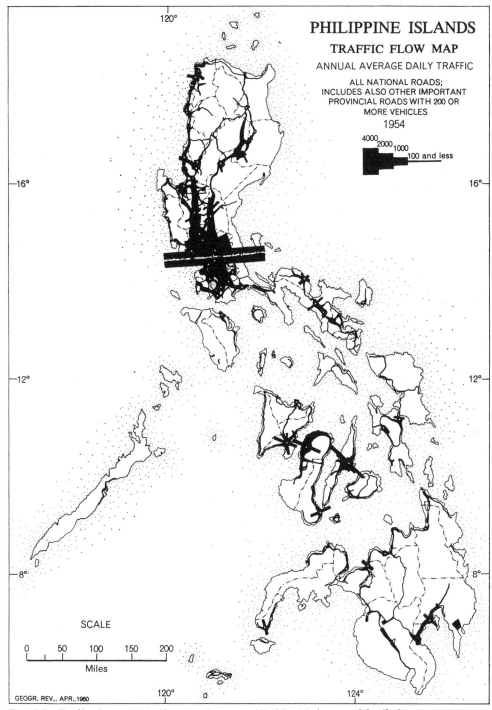

Fig. 8.2. Traffic flow, 1954. (Source: Bureau of Public Highways, Manila.)

THE HIERARCHY OF CENTERS
AND THEIR TRIBUTARY AREAS[5]

Five types of centers were recognized (fig. 8.3): *national center*: Manila (population 1,700,000); *interregional centers*: Cebu, Iloilo, Davao, and (partial) Zamboanga (population about 50,000–200,000); *major centers*: centers (33) of large trade areas, all but two of them provincial capitals; most have soft-drink warehouses and gasoline depots (population 10,000–40,000); *secondary centers*: centers (34) similar to major centers but less important (population 5,000–25,000); *minor centers*: small retail and social centers (126) (population 1,000–5,000).

National Center

Manila, the national center and the gateway to the Western urban world, is a typical primate city; as in most other underdeveloped areas, it is the one heavily commercialized center. Although it is not located at the geographic center of the country, Manila serves all the Philippines for many specialized central services and is as well the interregional center for a large immediate hinterland (fig. 8.3, inset). Manila far exceeds all other Philippine cities in volume of imports, manufacturing, government, education, communications, and a vast range of specialized social and administrative services. Its newspapers, to take but one example, penetrate all over the islands; in other centers the press is virtually nonexistent. It is the center of the airline network (fig. 8.4). The size of its population, nearly ten times larger than that of the next largest center, Cebu City, is a measure of its primacy. It has the strongest land connections to the largest land mass in the Philippines. Two railroads and several paved highways lead out from it. In domestic water transport, it is almost equal to Cebu City. Half of the petroleum imports come through its port, and half of the domestic consumption of lumber is shipped to its market (fig. 8.5). As is normally the case with cities, Manila's main commodity role is the distribution of imports. Raw-material exports are generally shipped directly abroad from coastal landings scattered throughout the islands, but some are handled through entrepots, including Manila.

"In 1956, I was quite optimistic about the development of a network of secondary cities; I don't know whether this is justified now. As a matter of fact, I suspect that Manila, the primate city, has continued to gain. This is a persistent problem in underdeveloped areas and many arguments, pro and con, have been advanced. Many have argued that in a relatively underdeveloped country it may be more economic to have growth in the primate city, although generally government policy (as opposed to government practice) and our desires are to try to get some secondary city or cities developed." [Letter to Donald Jones, 13 May 1974.]

"I have bucked up against the results of lack of good works in geography many times, most particularly when I served as a member of the Harvard Geography Committee. Instead of having the backing of a lot of good works to recommend to my colleagues, I felt almost alone in fighting the battle for geography. I still believe in geography and have spent much energy in fighting for it, but it is rather hard to fight honestly for something that has not produced. Furthermore it seems to me that our future obviously lies in the hands of the present students, who must become *better* than we are." [Letter to Glenn T. Trewartha, 19 March 1953.]

5. An extensive descriptive table of trade centers and tributary areas is available in mimeographed form upon request to the *Geographical Review*.

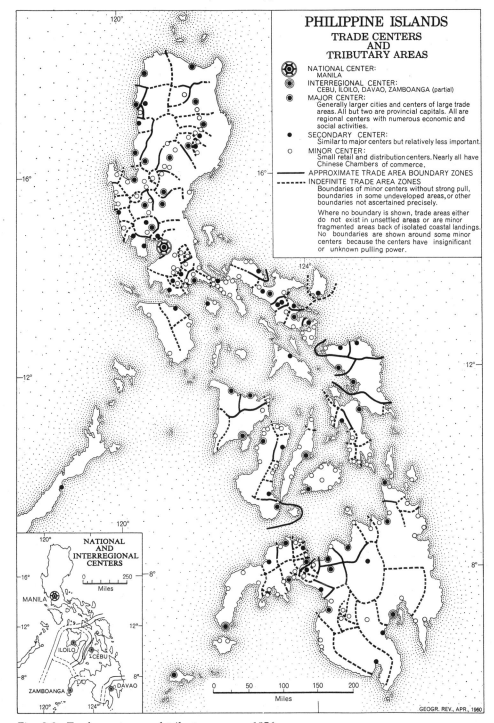

Fig. 8.3. Trade centers and tributary areas, 1956.

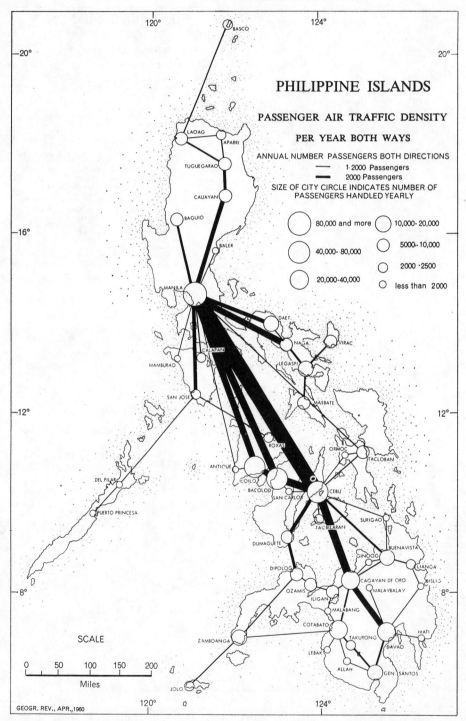

Fig. 8.4. Airline passenger flow, 1955. (Adapted from map by H. E. Robison, based on Philippine Air Lines data.)

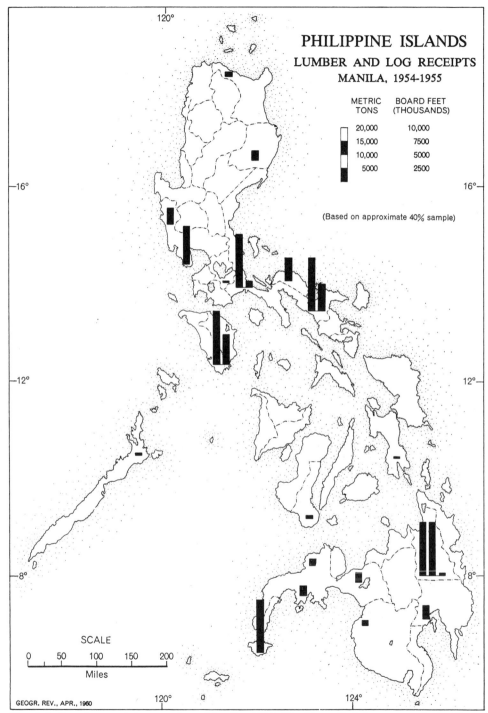

Fig. 8.5. Lumber and log receipts, Manila, 1954–55. (Compiled from records in files of Philippine Lumber Producers Association, which cover about 40 percent of shipments.)

Interregional Centers

Including Manila, five interregional centers serve as major subcenters in the Philippines (fig. 8.3). The other four are small, ranging from one-tenth down to one-fiftieth the size of Manila. These centers resemble small wholesale centers in the United States. They are the cities in which are located branches of firms serving all the Philippines for a multitude of economic activities.

Cebu City (population 180,000) serves as the center for most of the southern half of the Philippines. For a few functions it even competes with the other interregional centers in the south. Its hinterland, unlike Manila's, is reached almost entirely by water, the only means of connection with the numerous islands that constitute the hinterland. For this reason, it has a slightly greater domestic trade by water than Manila. However, its land trade on narrow Cebu Island is meager. Cebu City functions also as an entrepot, particularly for copra exports, for which it is the leading trade center.[6]

On the north the Cebu trading area meets the Manila area. The island of Samar is thus in many ways more closely tied to Manila than to Cebu. More passengers go by ship to Manila from Samar, but Cebu City is probably equally, if not more, important as a general supply source. Cebu collects more copra from Samar, and petroleum and many other products are sent from Cebu City to Samar. The island of Masbate, farther north, is also shared with Manila.

Iloilo (125,000), traditionally the third center, divides the central Philippine area with Cebu. The north-south alignment of the islands of Cebu and Negros creates a barrier to east-west movement by land and water and results in the curious, but logical, arrangement of two centers serving what might otherwise be served by one. The airline distance between these two interregional centers is less than a hundred miles, whereas the normal spacing for such centers in the Philippines is about three hundred miles. Iloilo faces further competition from the rise of Bacolod City on Negros Island, twenty miles to the east across Guimaras Strait. Formerly most of the sugar trade of rich Negros was transferred through Iloilo because it had the only deep, protected harbor; now most of the sugar is lightered directly to offshore ships on the Negros coast, in spite of the shallow water and the lack of suitable parts. Bacolod City (65,000) has become the center of the sugar area, partly because transport on excellent paved highways has replaced the older water transport to Iloilo.

6. F. L. Wernsted, "Cebu: Focus of Philippine Interisland Trade," *Economic Geography* 32 (1956):336–46.

Numerous central services have developed, and in some ways Bacolod, with the small but rich hinterland of northeastern Negros Island, is approaching interregional status.

It is probable that Manila is more of an interregional center than Iloilo for the north coast of the island of Panay, on which Iloilo is located; at least, Manila serves Kalibo and the new province of Aklan on the northwest. Roads cross the island, but they are poor, though a railroad runs from Iloilo to Roxas, on the north coast.

Davao (60,000), in southeastern Mindanao, has an isolated trade territory but serves as an interregional center, with road and water connections to surrounding areas. Zamboanga (25,000), in western Mindanao, serves partly as an interregional center. Its hinterland is reached almost entirely by water, since connecting roads have been completed only for short distances along the Zamboanga peninsula. The south-central coast of Mindanao, including Cotabato, is also probably loosely in Zamboanga's territory. Cotabato likewise has strong ties with Manila and Cebu City. However, passenger flow from Cotabato to Zamboanga is apparently slightly greater than that to Manila or Cebu.

Major, Secondary, and Minor Centers

The thirty-three major centers and thirty-four secondary ones can be considered together, since they are similar in most functions and generally differ only in size (fig. 8.3). These centers are important because they furnish the basic regional urban services to the country. They are small, however, and should not be confused with wholesale centers in the United States; rather, they resemble large county seats and retail centers, providing a wide range of economic, political, and social services.

The trade areas of most of the major and secondary centers are of the same order of size, and the centers are generally central in their respective territories. In developed and reasonably homogeneous territory spacing is rather regular (about forty airline miles apart), as would be expected, but the Philippines are so divided by water, mountains, and variation in population density that regular spacing is impossible over the greater part of the country. The Baguio trade area in northern Luzon is particularly eccentric, with Baguio in the southern corner, and is a special case, being composed of mountainous territory that normally would be a dividing zone. However, Baguio, an important resort center, is connected with its trade area by a remarkable long ridge road, which serves as the spine of the trade area (fig. 8.6).

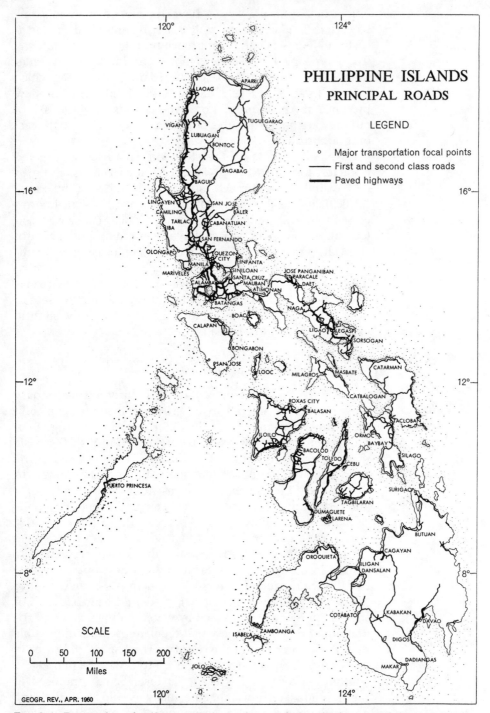

Fig. 8.6. Principal roads. (Source: First- and second-class highways, Philippine Studies Program, University of Chicago, and Bureau of Public Highways, Manila, ca. 1952; paved highways, author's observations, 1956.)

Most of the major and secondary centers are on the coasts, and most of them are ports. Their trade areas are thus generally coincident with their port hinterlands. However, many of the "coastal" cities are not directly on the coast but are a short distance back and have an outport (for example, Daet-Mercedes, Kalibo-New Washington, Roxas-Culasi) (figs. 8.7 and 8.8). This sort of arrangement is an unrecognized phenomenon characteristic of perhaps half of the coastal cities of the world, especially if the largest and best-known centers are excluded. In the Philippines, and to some degree elsewhere, the reasons for it apparently include hostility of the sea, both from pirates in the past and from nature, poor coastal sites, increasing size of vessels, which makes small upstream river ports unusable, and, perhaps most important generally, the fact that a city a short distance back from the coast is more centrally located to serve its hinterland in many local functions.

The areas that are particularly irregular and unorganized are the unsettled or sparsely settled, rough, mountainous regions along the northern and central east coast of Luzon, eastern Mindanao, south-central Mindanao, southern Negros Oriental, Palawan, and most of Mindoro. Some of these areas will develop, but difficult terrain and access are problems in most of them.

The recognized minor centers number 126. Undoubtedly others could be added, but the category gradually fades out to smaller and smaller centers. Practically all the minor centers have Chinese Chambers of Commerce but none of the other major indicators. Still smaller centers do not have even this function, and thus the Chinese Chambers of Commerce generally represent the cutoff point in defining minor centers. Traffic to most of these centers is so small that it did not register in the traffic-flow counts, and consequently it is impossible to delimit most of their trade areas.

Below the minor centers are more than 17,000 barrios, essentially rural farm settlements (figs. 8.9 and 8.10) inhabited by farmers who commute to their fields.[7] Most of the rural population of the Philippines live in agglomerated settlements, relatively few in isolated farmhouses. The barrio settlements have some rudimentary central services. About eighty percent have schools of the three to four lowest grades, the most widespread function, and the highest figure in the Orient outside Japan. Market day is generally held once a week for half a day, and a few *sari-sari* (small stores or stands) operate at all times. The most important barrios are the *poblaciones*, or central barrios, in the small clusters of

7. G. F. Rivera and R. T. McMillan, *The Rural Philippines: A Cooperative Project of the Philippine Council for United States Aid and the United States Mutual Security Agency* (Office of Information [MSA], Manila, 1952), p. 5.

Fig. 8.7. Roxas City, on the north coast of Panay, is a major trade center of about 16,000 population. It is situated on a river too small to accommodate ocean-going vessels. Road at upper left leads to Culasi, the outport for Roxas City.

Fig. 8.8. Culasi, the outport about five miles to the west of Roxas City, lies on deep water at the mouth of a lagoon.

Fig. 8.9. Farm villages in southern Luzon. These villages, strung along the roads in a *strassendorf* pattern, are close together because the land in rice terraces is so productive.

Fig. 8.10. Igorot farm villages in the mountains of northern Luzon. The village is surrounded by rice terraces watered by an ancient native irrigation system.

barrio settlements. Generally only the *poblaciones* have the higher grades of primary school.[8]

THE DEVELOPMENT OF SERVICE AREAS

As roads and commercialization increase in the Philippines, a rational framework of service areas is apparently developing: smaller centers and isolated coastal landings are giving way to major centers, which are extending their hinterlands by road (fig. 8.6.). An excellent example is found on the island of Samar, where a transisland road has recently been completed from Catbalogan, the capital, to the east coast. As a result, ship service to minor centers and coastal landings on the exposed east coast have been discontinued, and the whole eastern zone has become part of Catbalogan's hinterland. The same pattern will probably occur elsewhere. The truck and the auto are more efficient for local distribution than the ship. It is also sounder practice to concentrate port development in one major port for each island or major region than to try to maintain many small, poorly equipped ports. The traffic is generally insufficient to maintain more than one good port. In the meantime, the system is in a state of flux in many parts of the islands, and some areas are still virtually unsettled and unorganized.[9]

Along routes in many settled areas spacing is fairly regular and fits well the fundamental tenet of a central place system: the size and distance apart of centers are a function of the amount of productive land. This is exemplified in the rather regular spacing of centers north of Manila and into the Cagayan Valley of northern Luzon (fig. 8.3.), with their trade areas elongated at right angles to the road, as was to be expected.

In the Philippines five size classes of trade centers were established, a common and convenient breakdown all over the world. The first two, national and interregional centers, are clear, though the lower end of the second merges into the third. The third and fourth categories, however, major and secondary centers, might well be considered one, and only four classes would then be recognized. On the other hand, F. H. W. Green has demonstrated that freely developed bus services in England focus on fourth-order centers, which are probably similar to the Philippine third and fourth orders (major and secondary) combined.[10] One might well speculate that in an underdeveloped and poor country such as the Philippines there is no third-order center, or, rather, that all the centers are at least one grade lower than in rich, well-developed countries. Furthermore, in all primate-city coun-

8. Rivera and McMillan, *The Rural Philippines*, p. 147.

9. For an example of other effects of new roads on trade centers elsewhere, note the work of Professor Casas Torres and others in Spain as reviewed by Marvin W. Mikesell, "Market Centers of Northeastern Spain," *Geographical Review* 50:247–51.

10. F. H. W. Green, "Bus Services in the British Isles," *Geographical Review* 41 (1951):645–55.

tries, whether underdeveloped, like the Philippines, or well developed but small, like England, a second category is missing. Whether it will develop as the country develops is unknown. In the Philippines, no evidence points in this direction at present.

The number of centers in each hierarchical class, not surprisingly, does not follow the theoretical framework of either Christaller or Lösch (k=3, k=4, or k=7, which means that the number of equal size cities in lower size classes increases regularly by three, four, or seven times).[11] In the Philippines there are a greater number of major and secondary centers than the models call for, a fact that may well reflect the insularity of the country and a naturally fragmented set of hinterlands as well as the changing transportation system and poor measures of size and numbers. The number of Philippine centers in each class compared with representative systems of Christaller and Lösch is shown in table 8.1.

The Philippine distribution most closely approaches k=4. If seventy-three third-order centers are taken (by combining major and secondary into one category), the gift is closest to k=7; however, this is a questionable procedure. The k=4 arrangement, according to Christaller, is the most efficient for transportation, and this may be of some significance. In the final analysis, of course, elements of all three systems, in various parts of the islands, are to be expected, if underlying conditions permit; a close fit theoretically to any one should not be expected.

TABLE 8.1

COMPARISON OF NUMBER OF PHILIPPINE CENTERS WITH OTHER REPRESENTATIVE HIERARCHIES

| Center Size | Philippines | | K=3* | K=4* | K=7* |
	Actual Number	Cumulative			
1 (National)	1	1	1	1	1
2 (Interregional)	4	5	3	4	7
3 (Major)	33 } 67	38 } 73[a]	9 } 27[b]	16 } 64[b]	49
4 (Secondary)	34	72 } 199[a]	27 } 54[b]	64 } 192[b]	343
5 (Minor)	126	242	81	256	2401

* Cumulative.

[a] Arrived at by adding 67 (actual number of 3rd- and 4th-order Philippine centers) to 6 to produce new cumulative totals of 73 for 3rd-order centers and 199 for 4th-order centers.

[b] For comparison only with the Philippines, the same procedure as in the preceding footnote has been followed, but this has no real validity.

11. August Lösch, *The Economics of Location*. Translated from the 2d rev. edit. [1943] by W. H. Woglom, with the assistance of W. F. Stolper (New Haven, 1954); Walter Christaller, *Die zentralen Orte in Süddeutschland* (Jena, 1933). See also Edward Ullman, "A Theory of Location for Cities," *American Journal of Sociology* 46 (1940–1941):853–64.

Amenities as a Factor in Regional Growth   CHAPTER 9

For the first time in the world's history pleasant living conditions—amenities—instead of more narrowly defined economic advantages are becoming the sparks that generate significant population increase, particularly in the United States.* In spite of the handicaps of remote location and economic isolation, the fastest growing states are California, Arizona, and Florida. The new "frontier" of America is thus a frontier of comfort, in contrast with the traditional frontier of hardship. Treating this pull of amenities puts me, I realize, in the company of promoters and the traditionally uninformed, but if I make myself one with them, it is for new and valid reasons.

MOTIVATION OF MIGRATION

Modern writers on migration apparently agree that, except for forced shifts, economic opportunity is its motivating force. In 1934 the distinguished climatologist C. Warren Thornthwaite, commenting on California's phenomenal growth from 1920 to 1930, concluded, "Since the movement is abnormal in most respects, it is inconceivable that it will continue."[1] In 1938, Rupert Vance, distinguished sociologist, said: "On the basis of the exploitation of undeveloped resources of soil, minerals, forestry or water power, there can be expected no revival of the great westward migrations of the past."[2] In 1941, Margaret L. Bright and Dorothy S. Thomas[3] noted that California migration before 1930 far exceeded expectations based on laws of migration such as Stouffer's "intervening opportunity" and concluded: "We are of the opinion that an important part of the migration to California has been of hedonistic rather than a primarily economic character and has been motivated more by climate and legend than by superior job opportunities."

*Adapted from a paper of the same title read at the Seventeenth International Geographical Congress in Washington, D.C., August, 1952. Thanks are due to the Office of Naval Research for support of research for parts of this paper and to scholars who have read earlier versions, including Professors Thomas R. Smith, University of Kansas, Harry Bailey, University of California at Los Angeles, James Parsons, University of California (Berkeley), and Walter Isard, Massachusetts Institute of Technology; members of the Western Regional Economic Analysis Committee of the Social Science Research Council; and others noted subsequently. The author alone assumes responsibility for the contents.

Ullman is known for his clarification of the role of amenities in regional growth, a subject on which he published widely, reaching diverse audiences (Geographical Review, January 1954, pp. 110–32; Arizona Business and Economic Review, April 1954; Saturday Evening Post, 5 June 1954, p. 12). The basic material was reprinted in the Bobbs-Merrill Reprint Series in Geography in 1968, and in Economic Geography: Selected Readings, ed. Fred E. Dohrs and Lawrence M. Sommers (New York: Thomas Y. Crowell, 1970), pp. 286–99.

The significance of amenities as a major factor in regional growth was an idea pioneered by Ullman that has become a fundamental feature of economic geography texts today. As is evident from his comments and the article included here, amenities

1. C. W. Thornthwaite, assisted by H. I. Slentz, "Internal Migration in the United States," *Study of Population Redistribution Bulletin No. 1*, Wharton School of Finance and Commerce, University of Pennsylvania, Philadelphia, 1934, p. 18.

2. R. B. Vance, "Research Memorandum on Population Redistribution within the United States," *Social Science Research Council Bulletin* 42, 1938, pp. 85–110, reference on p. 92. Vance states further: "The one other chance for continued westward movement is industrialization; and there the Pacific Coast may reasonably expect to supply more of its own needs but not to dismantle the country's prevailing industrial distribution. Nor is it held likely that the California movement will continue at its former rate" (p. 92).

3. M. L. Bright and D. S. Thomas, "Interstate Migration and Intervening Opportunities," *American Sociological Review* 6 (1941):773–83, reference on p. 778.

become important in regional growth not simply because of site, as many supposed, but because of the loosening restraints of situation. It is only after transportation and its situational aspects are brought into play, that local, or site, features can become significant. Thus, Ullman recognized that regional development had changed from the nineteenth-century von Thünen model, in which transportation was so poor that it overpowered local site amenities and land use was determined almost entirely by proximity to market, to a situation emerging today in which transportation has so much improved that such site features as climate can emerge as major considerations in local development.

The article reprinted here appeared in the Geographical Review *(January 1954).*

4. Sixteenth Census of the United States, 1940, "Population: Internal Migration, 1935 to 1940—Color and Sex of Migrants" (U.S. Dept. of Commerce, Bureau of the Census, 1943), p. 18.

5. *Current Population Reports,* series P-25, Population Estimates, no. 12, U.S. Bureau of the Census, 9 August 1948, p. 9. In all three periods California's net in-migration was the largest in the country, several times larger than that of the nearest competitor from 1940 to 1945. From 1945 to 1947, however, California (108,000) was barely ahead of Illinois (106,000) and New York (83,000).

6. *Wall Street Journal,* 5 August 1952.

7. Calculated from "Historical Statistics of the United States, 1789–1945" (U.S. Bureau of the Census, 1949), pp. 33 and 38. See also *International Migrations,* ed. W. F. Willcox, Interpretations, vol. 2. Publications of the National Bureau of Economic Research, no. 18 (New York, 1931), p. 88.

8. K. J. Pelzer, *Population and Land Utilization* (An Economic Survey of the Pacific Area, Part I), Institute of Pacific Relations (New York, 1941), p. 26.

9. Calculated from Frank Lorimer,

As everyone knows, the influx into California was greatest between 1940 and 1950; even from 1935 to 1940 it was greater than to any other part of the country, averaging more than 175,000 a year, with a net of more than 130,000.[4] *Net* immigration from 1 April 1940 to 30 June 1941, was more than 300,000; during the war the net rose to an annual average of 422,000, and for the two years after the war the annual average was more than 100,000.[5] Immediately after the war some migrants started east, but within months the flow reversed, and under peacetime conditions the in-migration resumed on a large scale. A relatively high rate of unemployment resulted until Korean war orders took up the slack.

Figures for later years are not available, but reports from movers indicate that 1952 was their boom year for intercity moves, exceeding the previous largest year, 1951, by 20 percent, with heaviest moves toward the west and south. California, Texas, and Florida led, with New Mexico and Arizona also as fairly large net gainers.[6]

California migration is large-scale even in world terms. Apparently the largest previous migration in America was to the Prairie Provinces of Canada in the early 1900s, which reached a peak of 200,000 in one or two years. The greatest *net* immigration into the whole of the United States apparently was about 800,000 in 1910 and 1913.[7] Chinese emigration to Manchuria reportedly exceeded 1,000,000 a year from 1927 to 1929, though the net ranged only from about 400,000 to 800,000.[8] Net annual average in-migration to *all* of Asiatic Russia between 1926 and 1939 apparently was about 270,000.[9]

All these other great migrations were induced primarily by economic opportunity. California, on the contrary, received the first large-scale in-migration to be drawn by the lure of a pleasant climate, though other factors have played a role—including war, which caused airplane production to boom and enabled thousands of servicemen to see California for the first time. War, however, in part appears to be one of the shocks precipitating changes due to other long-range trends.

In the United States as a whole, the greatest changes in distribution of population between 1940 and 1950 were: (1) the suburban flight, a 35 percent increase in suburban population, as compared with 13 percent inside city limits and 6 percent elsewhere;[10] and (2) the growth of California, with a 53 percent increase, Arizona, 50 percent, and Florida, 46 percent, followed by Oregon, 39 percent, and Washington, 37 percent.[11] Undoubtedly the suburban flight has a large element of amenity seeking behind it, and was made possible by the automobile. This type of local migration will not be fur-

ther considered, though it reinforces some of my later conten-
tions on regional migration.

CLIMATE AS AN AMENITY

People have their violent preferences and prejudices, starting in a majority of cases with the conviction that where one
was born and lives is the best place in the world, no matter
how forsaken a hole it may appear to an outsider. Nevertheless, a substantial minority (millions in a country the size of
the United States) have other ideas. The first requisite is a
pleasant outdoor climate. The best criterion I can think of is
an outdoor climate similar to the climate maintained inside
our houses—a temperature of about 70°F. and no rain. Note
that emphasis is on a pleasant climate, a "nice" climate, not
necessarily one that drives men to the greatest physical or
mental efficiency, as defined by Huntington, Toynbee, Markham,[12] and others. Since the majority of Americans live in
the Northeast, in a colder, long-winter climate, this means
attraction of warmer climates.

A rough but objective ranking of the regions of the United
States in climatic pull puts coastal Southern California and
the climatically somewhat similar protected coastal areas of
Central and Northern California alone in Class I. This, the
only Mediterranean-type climate in America, has relatively
warm winters and relatively cool summers, coupled with low
rainfall and abundant sunshine. Some might also include a
small strip along the lower east coast of Florida, primarily
because of its winter pull. This area, the "Florida Tropics," as
demonstrated by Carson,[13] has a unique combination of
warm and relatively sunny winters, and a summer without
excessively high temperatures because of ocean exposure and
the cooling effect of winds. The most unpleasant feature
probably is the long length of the summer. This small area of
"tourist" climate is the one that has grown by far the most in
Florida, just as coastal Southern California has in California.

Class II areas will not be considered in detail, because of
limitations of space and variations in taste and local climates.
These areas might include a thin coastal strip along the Gulf
and South Atlantic (a winter resort for the North and a summer resort for the South), parts of Arizona and New Mexico,
protected parts of the remainder of the Pacific Coast, and
other local areas with more benign climates than their
neighbors, such as the Colorado Piedmont or Cape Cod.

Climate is probably the most important regional amenity,
because it can be combined with other amenities, especially

"The Population of the Soviet Union: History and Prospects," *League
of Nations Publications, II: Economic
and Financial, 1946.II.A.3.* (Geneva,
1946), p. 164.

10. Increase in population of
standard metropolitan areas as a
whole was 21.2 percent, as compared with 5.7 percent for the remainder of the country (Census of
Population: series PC-3, no. 3, U.S.
Bureau of the Census, 5 November
1950). The fact that most of the 21.2
percent increase took place outside
central cities is the basis for describing it as a "suburban flight." This,
however, probably overstates the
case, inasmuch as central-city boundaries were not expanded much and
consequently the increase in metropolitan population had to take
place outside the city (cf. Svend
Riemer, "Escape into Decentralization?" *Land Economics* 24 (1948):40–
48). Nevertheless, population has actually decreased in places near the
core of many large cities and has remained static elsewhere in some central cities, not increasing the national
rate. What is happening in cities appears to fit the amenity hypothesis;
Riemer notes that the trend "is not
toward 'decentralization' but toward
better residential districts which—
due to unfortunate circumstances—
are available only at the outskirts of
the city, and accessible only at the
cost of long commuting distances"
(p. 41).

11. *Statistical Abstract of the United
States: 1951*, U.S. Bureau of the Census, p. 31. Nevada had a 45 percent
increase, but in absolute numbers
this represented an increase of only
50,000, too small to be statistically
significant. Arizona's increase of 50
percent is in somewhat the same category, since it represents only
250,000 persons. In absolute gain
(rounded figures) California was far
in the lead with about 3,600,000 followed by New York, 1,300,000;
Michigan, 1,100,000; Ohio, 1,000,-
000; and Florida, 900,000 (slightly
more than Illinois). If one considers
estimated increases of civilian population only, up to 1 July 1951, the
ranking for the 11½-year period
changes slightly: Arizona, 58 percent; California, 54; Florida, 52;
Nevada, 51; Oregon, 42; Washing-

"I include a preliminary draft of a paper on amenities and regional growth, which attempts to show that the changes in America now make for a redistribution of population based, for the first time in the world's history, on the 'niceness' of places." [Letter to Charles E. Odegaard, president of the University of Washington, 17 July 1952]

———

ton, 35 (*Current Population Reports,* series P-25, no. 62, U.S. Bureau of the Census, 24 August 1952). Estimates of net in-migration gain, 1940–1950, are: California, 38 percent; Nevada, 31; Florida, 30; Arizona, 28; Oregon, 26; Washington, 23 (ibid., no. 72, May 1953). Percentage increases for the period 1 April 1950 to 1 July 1952, for the fastest-growing states are: Arizona, 15; Nevada, 12; Florida, 12; Maryland, 8 (suburban spillover from Washington, D.C.?); Colorado, 8; California, 8. Absolute increases in the same period by rank are: California, 804,000; Texas, 447,000; New York, 348,000; Michigan, 337,000; Florida, 329,000 (ibid., no. 70, 24 March 1953).

12. Ellsworth Huntington, *Civilization and Climate* (3d ed., New Haven: Yale University Press, 1924); idem., *Mainsprings of Civilization* (New York and London: J. Wiley, 1945); S. F. Markham: *Climate and the Energy of Nations* (London, New York, Toronto: Oxford University Press, 1944).

13. R. B. Carson, "The Florida Tropics," *Economic Geography* 27 (1951):321–39.

14. C. A. Mills, *Climate Makes the Man* (New York and London: Harper, 1942), p. 289 and *passim*.

15. C.-E. A. Winslow and L. P. Herrington, *Temperature and Human Life* (Princeton, N.J., 1949), pp. 254–55.

16. P. H. Parrish, "Refugees from the Middle West," in *Northwest Harvest*, ed. V. L. O. Chittick (Writers' Conference on the Northwest, Portland, Oregon, 1946) (New York, 1948), p. 54.

within the continental United States, where there is a fairly even spread of culture, education, sanitation, and creature comforts of all sorts. The best scenery and bathing in the world are useless unless one can get out in them. Furthermore, climate has an important effect on the health of many sufferers, warm, dry regions outside storm-track zones (southwestern United States), according to Mills,[14] apparently doing the greatest good for the greatest number of ailments, though Winslow and Herrington state: "Thus, considering all the evidence at hand, we can only predicate with certainty that extremes of heat and cold are definitely harmful; and that even moderately hot conditions increase susceptibility to intestinal diseases, and moderately cold conditions increase susceptibility to respiratory diseases."[15]

Other amenities, however, do exert a pull; mountains and beaches, hunting, fishing, and other sports, beautiful New England towns, all come to mind. Even if the Great Plains had a near-perfect climate, they probably would not lure as many people as the same climate in a region with mountains and water. This is a subjective matter: some people seem to like flat country, but most residents of the Pacific Coast (since most of them are refugees from the Middle West) would probably gain solace from the fact "that no matter what may happen to them, no matter what their lot in life may be, they do not live in Kansas."[16]

The rest of the world will not be considered except to note the two types of areas with "ideal" climates: (1) parts of the other "Mediterranean" climatic regions, and some ocean-tempered tradewind islands such as Hawaii; and (2), potentially the best in the world, high altitudes in low latitudes—parts of tablelands in Latin America and other tropical mountain zones.

Outside the United States, the population generally is less wealthy and foot-loose, and no growth related to climate comparable with that in California has taken place, with two possible exceptions, both in countries somewhat similar to the United States in wealth, and each with a "nicer" place within its borders to go to. In Canada, Vancouver is the most rapidly growing city. The Vancouver region is no California, but compared with the rest of Canada, it has the best climate and scenery as well as other attractions. In France, Nice, on the Riviera, was the fastest growing city before the war.

If, for the first time in the world's history, the population of "nice" areas in some countries is growing more rapidly than that in the remainder of those countries, the fundamental question is: What has happened to the economy to make this possible? Following are some factors, many of which need further research to establish their quantitative contribution.

RETIREMENT AND TOURIST FACTORS

The growth of early, paid retirement, coupled with longer life expectancy for the population as a whole, is one factor. Even in the high-birth-rate period from 1940 to 1950, the number of persons over sixty-five years of age increased from 6.9 percent of the total United States population to 8.2 percent. Industrial unions have been obtaining retirement provisions so generally that the number of retired workers will increase enormously in the future; so also will the number aided from expanding Social Security.[17] Furthermore, old people seem to like a warmer climate, and, as noted, most of the workers now live in the colder climates of the Northeast or Midwest. The net effect of the removal of a number of these people to places such as Florida or California is a subsidy from one region to another.[18] However, only a portion of amenity-induced growth can be attributed to this factor, since people over sixty-five are merely a small part of the United States population. The National Planning Association estimates that in twenty-five years they will number 20 million, of whom 14 million will not be working.[19]

Related to retirement is the well-known growth of the tourist industry, partly a response to the spreading practice of paid vacations, even for industrial workers. In 1940, only one-fourth of all labor contracts called for paid vacations; now almost all do. Altogether, 42 million workers are eligible, many of them for increasingly long vacations, up to three weeks or more.[20] Florida and California derive substantial incomes from this trade, as do amenity regions closer to home for the majority of workers, such as New England.

INCREASE IN FOOT-LOOSE WORKERS

Increase in number of foot-loose workers is related to war production, but particularly to the long-range trends established by Colin Clark.[21] A decrease in the number of primary workers (agriculture, fishing, and forestry), because of increasing mechanization and agricultural efficiency; a static or, in some cases, declining level of secondary workers (manufacturing, mining, and construction) except perhaps in wartime; and a great increase in tertiary employment (trade and services).[22]

Logically, the increase in tertiary employment should occur

17. "By mid-1950 practically every major union in the country . . . had to some extent negotiated pension or health and welfare programs" (E. K. Rowe, "Employee-Benefit Plans under Collective Bargaining, Mid-1950," *U.S. Department of Labor, Bureau of Labor Statistics Bulletin* 1017, 1951 [reprinted from *Monthly Labor Review* 72 (1951):156–62]). At least seven million industrial workers now have pension plans, most of them only a few years old; government old-age insurance is also increasing. However, as a counterbalance, the effects of inflation have cut into these benefits enormously and have also reduced the return from savings. Lower interest rates have had the same result (*Proceedings, Governors' Conference on the Problems of the Aging*, Sacramento, California, 15–16 October 1951, p. 282). Note, as a further counterbalance, the obvious fact that old people do not live as long as younger people and hence a given number are not as long-lasting a gain in population.

18. When wealthy people are involved, this may introduce a large amount of capital and start off a chain reaction. Carey McWilliams (*California: The Great Exception* [New York: Current Books, 1949], pp. 257 and 260), notes the effect of Pasadena retired millionaires in building and endowing Mt. Wilson Observatory and the California Institute of Technology.

19. "The Leisured Masses," *Business Week*, 12 September 1953, p. 146. See also *Problems of America's Aging Population*, T. L. Smith, ed., *Univ. of Florida, Inst. of Gerontol. Series* 1 (1951):15 ff., for evidence of migration of the aged to California and Florida.

20. "The Leisured Masses," p. 146.

21. Colin Clark, *The Conditions of Economic Progress* (London, 1940).

22. In 1850 primary employment amounted to 65 percent, secondary to 18 percent, and tertiary to 18 percent; in 1920, the percentages were 27, 33, and 40 respectively; and by 1950, 17, 34, and 49 respectively (1950 figures calculated from 1950 Census of Population: Preliminary Reports, series PC-7, no. 2, U.S. Bureau of the Census, April 1951,

". . . this deals with regional shift, primarily in response to natural amenities about which the planner can do little. However, I feel that the moral is the same for man-made amenities . . ." [Letter to Fred J. Abendroth, 31 August 1953]

"I can do no wrong in California, Arizona and Florida, but probably in the rest of the country, my name is mud." [Letter to Grady Clay, 30 December 1954]

----

pp. 31–33; figures for 1920 and 1850 taken from P. K. Whelpton, "Occupational Groups in the United States," *Journal American Statistical Association*, 21 (1926):340.

23. Harvey Wish, *Contemporary America* (New York, 1945), p. 31.

24. Address, "The Los Angeles Opportunity," by P. H. Willis, general advertising manager, Carnation Milk Company, to the Advertising Club of Los Angeles, 7 September 1948. According to *Fortune* ("Industrial Los Angeles," June 1949, p. 154), Vice-President Alfred M. Ghormley of Carnation noted that Los Angeles was the company's largest single market and was stimulating to management because it generated so many new grocery techniques, but "the most important factor" was that "while money means a lot to all of us, it certainly is not everything today, and we felt that we could attract better executive material if we could bring some of these men into a climate they might enjoy more."

25. J. J. Parsons, California Manufacturing, *Geographical Review* 39 (1949):229–41, reference on p. 240.

26. W. G. Cunningham, *The Aircraft Industry* (Los Angeles, 1951), p. 198.

27. L. W. Casaday, "Tucson as a Location for Small Industry," *University of Arizona, Bureau of Business Research Special Studies No. 4*, 1952, p. 23. The other quotations in this paragraph are from the same source, pp. 23 and 24.

in areas of primary and secondary employment; up to now this paper has established only an increased base of retired people and tourists to support this increased number of tertiary workers in benign areas. However, it seems reasonable that a growing but unknown number of tertiary workers are also nationally foot-loose. Many specialized services can meet the needs of a national market from anywhere in the country, such as the movie, radio, and TV industry in California. The cinema appreciates the same climate and scenery as humans (as also, to a certain extent, does California and Florida agriculture). Clear weather for shooting pictures, particularly in the industry's initial outdoor period, plus a variety of scenery, was a factor in locating the motion-picture industry in Hollywood, along with the specific flight of independents from business troubles in New York.[23]

In the business world also, there are indications of at least a partial effect of amenities on the location of activities, though whether they overbalance the presumed agglomerative benefits of an eastern, closer-to-market headquarters is unknown. The Carnation Milk Company, for example, has recently centralized its headquarters in Los Angeles. Part of the reason given (other than the need for centralizing operations and the relatively important market position of the company in the West) was better living conditions; another part was the professed "dynamic" quality of Los Angeles. National control from Los Angeles is considered feasible now because of the speed of traveling to, or communicating with, the whole country.[24]

In industry, evidence, again not yet as quantitative as desirable, indicates some pull of amenities. High-value products such as calculating machines or advanced electronic products can afford shipping costs all over the country from California. On the other hand, some companies producing a bulkier product that started in California have moved east to get efficient national distribution.[25] Likewise, some other industries benefit somewhat from a benign climate just as motion pictures do. This is one of the location factors for the otherwise mobile, somewhat outdoor industry of airframe assembly, the largest industry in California.[26]

In assessing the pull of amenities on foot-loose industries one runs into a reluctance of executives to admit that personal comfort considerations motivate them, "seeming to feel that the location of the firm ought to be justified on more objective grounds."[27] This reluctance, noted in Arizona, seemed to prevail also in the aircraft industry in California, according to personal conversation I have had with Glenn Cunningham, an authority on location of the industry, though he had no way of proving the point. Nevertheless, for thirty-four small industries studied by Casaday in Tucson, Arizona, climate was found to be the overwhelming attraction, not only to executives, but even more to labor, because of its favorable effect on labor availability, satisfaction, and efficiency. A representative comment of those interviewed is as follows: ". . . a shortage of labor will never develop here. Thousands of families in the east and midwest who have health problems or who have always wanted to move to Arizona would come at the drop of a hat if they were sure of steady employment and adequate housing. At the worst, a little direct advertising in the eastern part of the country would solve any labor supply problem Tucson is likely to have." Still another comment refers to a rapidly growing, relatively foot-loose phase of industry: "It would take a helluva lot to get me to leave Tucson but even if I did have to go back east I would see to it that the laboratory operations remain right here. There couldn't be a better location for that type of work."

MARKET-ORIENTATION FACTOR

The shift of industry to greater market orientation has apparently overbalanced the movement to raw materials in recent years and means a larger amount of industry supported by the increased population of the newly expanding areas.[28] Much of California's postwar manufacturing growth is market-oriented. Three long-range factors help explain the shift: changes in technology, in transportation, and in economies of scale.

In heavy industry, for example, economies in use of fuel have reduced raw-material requirements per ton of finished products;[29] this, along with other factors, has made it possible for Kaiser to produce steel in Fontana for the California market even though coal has to move more than 500 miles by rail from Utah. Apparently, also, the cost of transporting bulky raw materials, which can be handled mechanically in volume, has decreased more than that of shipping finished products requiring more hand labor, an increasing cost item. Thus an increasing percentage of fuel requirements are obtained from oil or gas, much of which now moves by pipeline longer distances than coal could afford to move by rail.

Finally, as a market increases in size, new economies of scale are possible and an additional number of new specialties can be supported on the increased base established. This has happened for manufacturing to some extent in California.

"Seriously, I don't maintain that amenities are the main reasons for development of regions, but they are an increasing factor; in any case, I have never written anything that gave me more fun to write. The article is being reprinted in whole or in part in Arizona and in Oregon. They all want to get on the bandwagon. One friend of mine said that I had hit upon a new device in scholarship—to take the obvious and make it learned rather than the reverse. The trouble with this method is that if one then tries to make a popular version of such an article, one is reduced to nothing." [Letter to Charles Upson Clark, 10 May 1954]

"I have been racking my brain for a better title for my paper and after rejecting several, the one that may seem best is 'A New Force in American Internal Migration.' (If you desire, you can put as the first word to this title, 'Amenities' and put a colon after it.) Frankly, I can't think of a synonym for amenities. I suppose I could use some sort of frivolous title as 'Hedonism and Regional Development,' or 'The Effect of the American as Sybarite,' or even 'Go West, Old Man.'" [Letter to Charles L. Hamman, 5 October 1954]

28. Even in the South; note the conclusions in G. E. McLaughlin and Stefan Robock, "Why Industry Moves South," *National Planning Association Committee of the South* report no. 3, 1949. About 45 percent of the new plants (and a larger percentage of employment) moved to serve southern markets, 30 percent to raw materials, and 25 percent to cheap labor (pp. 26–27).

29. Cf. Walter Isard, "Some Locational Factors in the Iron and Steel Industry since the Early Nineteenth Century," *Journal of Political Economy* 56 (1948):203–17.

"I've had only one good investment hunch from geography lately. After writing the Amenities article last year I became convinced that air conditioning was due for a boom. Then I read an article on Carrier Corp. in *Fortune* and decided that this publicity alone might help boost the stock. However, I did not have a dime to invest but I was so convinced that I decided to borrow $500 from the bank. On the way over to the bank my innate scientific caution deterred me and I did nothing. Shortly thereafter the stock market broke, but Carrier at least remained firm at 40 and finally in the last 3–4 months the stock has climbed steadily to about 60 where it is now, too high to buy on borrowed money!" [Letter to Richard U. Light, 19 March 1954]

30. In 1929, 6.3 percent of California's population was employed in manufacturing, 7.9 percent in the United States as a whole; in 1947 the percentages were 6.8 and 10.0, respectively (1947 Census of Manufactures, vol. 1, U.S. Bureau of the Census, 1950, pp. 35 and 39).

31. We do not yet know at what thresholds of population various increases in industry are likely to occur, except in general terms. In practice, lag and speculaton are involved, as well as absorption pricing policies of plants elsewhere in the country, and other factors.

32. *Business Week*, 7 March 1953. Some authorities indicate that costs for household air conditioning can be as low as $12 a month throughout the year to cover both operation and amortization. In a dry area such as southern Arizona simple evaporation systems, which work well except during infrequent humid periods, are even cheaper (personal communication from A. W. Wilson, University of Arizona, Tucson).

33. "The Changed America," *Business Week*, 6 June 1953, p. 112.

The state has had a real increase in percentage of population employed in manufacturing. However, as compared with national growth, manufacturing has not expanded as rapidly in California as population has. Thus the deviation of the percentage of California population employed in manufacturing from the national percentage was −1.6 in 1929 and −3.2 in 1947.[30] On the basis of these trends, economies of scale, both external (regional) and internal (single industry or plant), apparently have abundant scope for still greater application in the future.[31]

OTHER NEW FACTORS

Still other new factors, mostly social but partly technological, bear on the thesis that pleasanter places are due for an increase in population.

1. The greatly increased mobility of the American people, because of universal auto ownership and good roads, makes transcontinental moves reasonably commonplace and permits Americans to discover amenable regions during longer vacations.

2. As more people settle in pleasant areas, they themselves will exert an agglomerative pull, bringing in still more newcomers. As is well known, firsthand reports from friends and relatives are one of the strongest means of advertising for immigrants.

3. One of the results of these two factors is to bring to light a minor economic incentive to live in warmer climates; the lower cost of fuel, housing, and some other items.

4. The prospects for widespread air conditioning will make warm regions more attractive, especially in the United States, where income levels will be sufficient to cover costs.[32] Theoretically, this should relatively favor Florida and Arizona, with warm summers, rather than coastal California, with cool summers.

5. The present high birth rate is resulting in larger families and thus creating more mothers who wish they could let their youngsters run outdoors in winter and hope (probably in vain) that they would thereby escape the high and gloomy incidence of winter colds and flu. Anyone who has spent a winter in New England or the Midwest must have heard mothers wish that they could move to California or Florida (many do, probably a reflection of our matriarchal society!). As many have noted, "the really fundamental economic decisions are made in bedrooms not board rooms."[33]

6. In our society today the conviction apparently has

grown up, along with heavy taxation, that it is difficult to make a lot of money and consequently one might as well enjoy life—the reverse of the earlier emphasis on the hereafter. A pleasant place to live is given more consideration, other things being equal. Likewise, "mass leisure" is now a feature of the United States. Reduction in the industrial work week from sixty-four hours in 1860 to an average of forty-two in 1930 (and a further drop since then) has given the worker more leisure time to appreciate outdoor and other amenities.[34]

Most state planning and development agencies, and public utilities, recognize the lure of amenities, some of them with reason. Officials of a large utility in Chicago have told me that one of their biggest problems in luring industry to Chicago is the unwillingness of executives to live in the city. As a result the company has gone all out in advertising the presumed cultural and recreational advantages of its city. North Carolina advertises, "There is profit in pleasure"; New Hampshire, "There's a Plus in every pay envelope"; Colorado offers, as a minor inducement, the "magic" of the Colorado climate; and finally, British Columbia advertises itself as the "California of Canada"![35]

Underlying this thesis is the apparent hedonistic goal of the American people and of much of the rest of mankind; this goal may represent a new emphasis—but I doubt it—on tangible physical pleasure rather than on psychic pleasure from religion or prestige,[36] or a puritanical glow derived from hard work and acquisition of wealth, or a human stimulation from learning, culture, and the growth of the inner man.

Forces are also working in the opposite direction. These will not be treated here, nor do I feel that they counterbalance the forces allowing population to move to amenities. At most, they represent another pole pulling simultaneously, such as the growth of West Virginia and Texas, based on natural, chemical resources, along with that of California and Florida, based largely on climate.

CONCLUSION

Discovery of the spark starting regional development is crucial in view of the increasing number of service workers and industries dependent on the initial base. In singling out amenities for analysis, I have deliberately concentrated on a new, speculative force, whose workings are not yet understood and whose influence may be greater in the future. Nor is it my intention to explain in terms of a single cause so

34. Reuel Denney and David Riesman, "Leisure in Industrial America," in *Creating an Industrial Civilization: A Report on the Corning Conference* (held under the auspices of the American Council of Learned Societies and Corning Glass Works and edited by Eugene Staley and others [New York, 1952], pp. 245–46).

35. Others have also recognized the contribution of amenities, as witness the following statement: "Although, historically, migration within the United States was always associated with improved economic opportunities, the permanency of war migration to this region was apparently strongly influenced by the psychological factor of taste for the region and its climate and other preference imponderables. This unpredictable permanent increase of people in the Columbia Valley region has also been affected by the remarkable postwar expansion of commercial, industrial, and construction activity, which has easily absorbed the large number of migrants choosing to stay, as well as most of the returning veterans . . ." (Charles McKinley, "Uncle Sam in the Pacific Northwest," *University of California, Bureau of Business and Economic Research Publications*, 1952, p. 9). And still another, typical of an intermediate locality and bringing in home ties: "Consideration of climate and recreational facilities, or the fact that the owner has grown up in Michigan appear to determine in not a few cases the choice between a Michigan location and one in Ohio, Indiana, or Illinois" (James Morgan and Harold Guthrie, "What Michigan Manufacturers Think of Michigan," *Michigan Business Review* 3 (1951):18–20).

36. Note, for example, the increasing popularity of various house, garden, and living magazines of national and regional circulation. To mention one from the West, there is the popular and attractive *Sunset* magazine. Note also Ghormley's statement in footnote 24, above.

"The first and foremost fact that emerges from a study of change in population on a regional basis for the last ten to twenty-five years is the growth of California and Florida. Even from 1950–54 we have the same trend, with Florida growth at 25.9%, Arizona at 31.3% and California at 17.3%, with the rest of the states (with the exception of statistically insignificant Nevada) growing in very much lower percentages. For this reason I could not resist, several years ago, writing the enclosed 'Amenities as a Factor in Regional Growth.' The coincidence of population migration with what people consider to be nice climatic areas is too close and persistent to be explained away solely by mere coincidence." [Letter to Donald J. Bogue, 3 November 1954]

large a phenomenon as the growth of California or the recent migration of peoples.

Before definite conclusions can be drawn, further research and testing are required. Basic to this testing is analysis of the trends and probable degree of future foot-loose orientation of services and industries in terms of national location. This probably means a detailed and exhaustive analysis of growth of individual industrial products and services based on stage of process, which census classifications do not give.

As was noted before, the continental limits of the United States rather sharply contain the area within which amenities for Americans can operate on a large scale today, not only because of uniformly widespread culture and comfort, but mainly because linkages with the rest of the economy are easiest, and in many cases, only possible, within the continental United States. Because of the small area of subtropical climate within the country, California and Florida largely escape competition. Their amenity pulling power is reinforced by the relative uniqueness of their environment, which enables them to exert a pull even across half a continent. Thus Carson notes that southeastern Florida is probably the only "place on earth where middle latitude progressiveness meets the exuberance and livability of the tropics. . . . Inhabitants of other tropical regions may enjoy complete freedom from frost but are less likely to acquire a car, a mail order catalog, or a legacy of intellectual curiosity."[37] Migration to amenities and pleasanter climates appears to be one of the more reasonable results of what the economist Galbraith[38] calls the "unseemly economics of opulence," a more rewarding way of spending effort than in advertising cigarettes or degrading flour and then re-enriching it to make bread, or any of the countless other ways in which money is thrown around in our wealthy economy. Even Aristotle noted, "Men seek after a better notion of riches . . . than the mere accumulation of coin and they are right."

It looks as if America, given half a chance, might become a nation of sybarites. We now have this half chance. Oscar Handlin[39] observes, for example, that the American laborer who once hesitated to risk merely shifting from one factory to another is now willing "to move from one section of the country to another, confident he will anywhere find a demand for his services."

Even if our ends have not changed, our means have. And these changes seem to be just beginning so far as the predictable future is concerned. Even the unpredictable future indicates the same: Thus the climate of California and Florida

37. Carson, "The Florida Tropics," pp. 338–39. Note also Ackerman's conclusion that citrus fruits are grown in California and Florida, even though they are slightly colder than the optimum, because they are in the United States (E. A. Ackerman, "Influences of Climate on the Cultivation of Citrus Fruits," *Geographical Review* 28 (1938):289–302).

38. J. K. Galbraith, *American Capitalism* (Boston: Houghton Mifflin 1956).

39. Oscar Handlin, "Payroll Prosperity," *Atlantic Monthly* 191 (February 1953):31.

takes its place as a population magnet along with the coal of Pittsburgh and the soil of Iowa.[40]

Assuming that we have proved our case, what is the moral? There are at least three:

1. The amenity factor should be kept in mind in predicting future regional population and development; the predictor, however, is under special obligation to be objective, because most of mankind thinks his own region is best, and indeed may even be paid to think so.

2. Improvement of amenities of a city or region may actually pay off in the long run, something no planner has ever been able to prove. Here care should be taken not to kill the goose that lays the golden egg by crowding population and industry into a place in an unplanned and unpleasant manner and creating intolerable traffic, smog, and other conditions, as has happened in some cases, but need not.[41]

3. No matter how much man tries, he cannot compete with Nature in regard to some of the most important amenities, such as climate, though air conditioning will make warm climates more attractive in countries that can afford it, just as central house heating improved cold climates in the past.

"I too believe that manufacturing, since it is becoming more complicated, is getting farther removed from raw materials, and was glad to get your examples. For my paper on Amenities I particularly wanted to see whether truly footloose industries oriented neither to raw materials nor to markets were growing more rapidly than strictly market-oriented industries. By footloose I was thinking primarily of industries producing high value products which could be shipped all over, such as airplanes or electronics. . . ." [Letter to Chauncy D. Harris, 19 October 1953]

40. This is somewhat of a reversal of Ellsworth Huntington's optimum-climate hypothesis that the slightly cooler climates are more stimulating and therefore Northern Europe and the northern United States are the most "advanced" regions in the world. To me, this appears to be a reasoning after the fact, a fact for which coal and iron ore, strongly localized in these regions, and an earlier start were more important causes. (Huntington does, however, include coastal California in his optimum climate along with the Northeast; see his "Civilization and Climate" or "Mainsprings of Civilization".) Toynbee, in his challenge-and-response theory of history, advances a slightly different argument, though not limiting his environment to climate. McWilliams, "California: The Great Exception," employs Toynbee's thesis with a new twist in noting the different nature of California's environment and its consequent (?) stimulation to invention and growth. Markham, "Climate and the Energy of Nations," indicates much the same optimum climate as Huntington does and emphasizes the development of indoor heating based on coal in these areas as a way of overcoming the cold winters. He also notes that future air conditioning in wealthy countries (notably the

United States) may well minimize some of the handicaps of the warm climates.

41. Note the following opinion of a West Coast city manufacturer (quoted from the 24 January 1952 issue of *Direction Finding*, a service published by Industrial Survey Associates, 605 Market Street, San Francisco 5, California, and reprinted in *Area and Industrial Development Publications*, no. 18, Area Development Division, U.S. Department of Commerce, March 1952, p. 8):

"In response to your letter regarding an increase in our subscription to the chamber of commerce for the specific purpose of bringing in more industry, new people, and new payrolls, including greater quantities of sewage pollution on our beaches and rivers, greater congestion in living quarters and in our temporary school buildings, greater congestion of traffic and carbon dioxide gas pollution in our streets, and general destruction of our natural resources wherever increased population can spoil and destroy them—if these are the things you want, I certainly am not for them.

"Since we are in the manufacturing business we are naturally interested in payrolls, transportation, housing, schools, etc. We are considering plant expansion ourselves, but it certainly will not be in this city under present conditions. In fact, we are moving as far away from these congested conditions as we possibly can get, to a small town where we will have room to breathe and happy home-owning employees, far removed from the mess we have here. Until these conditions are cured, I am against bringing any more people to this area. . . ."

"I maintain that whether people think there are amenities in a place is the important consideration in their migration, regardless of the actual merits of the case. I suppose once people move to an amenity area, whether for valid reasons or not, they may tend to make it more amenable and thus it will sort of be perpetuated as a nicer place to go. . . . With the development of California and Florida in this country, work has to be created in those regions, and this seems to get done, although the population comes first and the work second." [Letter to E. W. Bilbert, 20 August 1954]

"It looks to me as though empirical evidence indicates that the northward movement has stopped and indeed has reversed. This applies, of course, only to a short span of years. However, New England is declining, and as the enclosed article shows, California, Arizona and Florida are by far the fastest growing states in the United States. The south also seems to be catching up with the rest of the nation. Moscow is now the capital of Russia, rather than Petrograd. Rome is the capital of Italy. Washington, D.C., is the capital of the United States. In Scotland the midlands are declining drastically. The south of France is growing rapidly. . . .

One new theory which I have in the enclosed paper on amenities is that as the standard of living rises, people are seeking out more benign climates just as they acquire television sets. Perfection of air-conditioning may make warm climates just as habitable as central heating did those of colder climates. A special case in favor of places like California, and to a lesser extent

Colorado Piedmont, are their cool summers. The American middle west is unbearably hot in the summer in comparison with the growing parts of California.

Personally I don't put much stock in Huntington's theory that variability of climate is desirable for mental progress. It would seem to me that a perfect climate the year around would be better than one that is too hot in the summer and too cold in the winter as is characteristic of most of the United States. Certainly the summers in Berkeley are much better to work in than those of Chicago, and I am sure more is accomplished. Places like California, it is true, have a great diurnal range of temperature.

Perhaps one reason for the supremacy of the cool climates such as those of New England, is that they are so miserable that no one wants to stop work to go outside. I used to make this point facetiously at Harvard. Huntington points out that libraries are rare in the south and fewer books are read. This is quite true, but isn't the reason possibly simply that kids when they grow up can play outdoors so much more that they don't have to entertain themselves by reading books and thus do not develop the habit?" [Letter to S. Colum Gilfillan, 24 September 1954]

Geographical Prediction, Regional Planning, and the Measure of Recreation Benefits in the Meramec Basin

Prediction, as opposed to guessing or mere description, is the goal of science. Logical understanding may be a sufficient objective for many, but by definition prediction is a contribution both to science, and by extension to society. This paper will present a method of *geographical* prediction based on comparative, or analog, and interaction approaches.[1] This procedure is somewhat revolutionary in geography, but can logically develop out of geographical thought. It will be applied to the measurement of recreation benefits, an intangible item difficult to measure. Geographical prediction will be demonstrated, therefore, by a real problem in a real place, using real data to produce a real answer.

Edward Ullman gave more detailed attention to the Meramec Basin in Missouri than perhaps any other area on earth. As director of the Meramec Basin Research Project at Washington University in St. Louis from 1959 to 1961, he was in charge of a comprehensive regional development study for this 3,000-square-mile area in the Ozarks. The purpose of the study was to find ways of encouraging development in the entire St. Louis trade area. Recreational reservoirs were found

THE MERAMEC BASIN STUDY AS A PREDICTION MODEL

The setting is the Meramec River basin in the American Middle West, a river basin about one hundred by fifty miles (160 by 80 kilometers) extending from the St. Louis suburbs in Missouri down into the Ozark Mountains. I was asked to conduct a study of the region for the purpose of general economic development, but with particular reference to water resources and flood protection—in other words, a specialized regional planning study.[2]

The area is heavily forested, rough, mostly plateau surface, and is one of the emptiest areas in terms of population in the eastern half of the United States, resembling the Adirondacks or the upper peninsula of Michigan; yet it is only fifty miles from a metropolitan area of two million persons. This immediately suggests, without any elaborate models or statistical preparation, a recreational relation between the metropolitan area and this wilderness area.

CHARACTER AND CONNECTIONS
OF THE REGION AS AN URBAN FIELD

Not only is the area sparsely populated, but it also exhibits in exaggerated fashion the typical gradient profile of change in character from city to country. Thus one passes from the central city, with its static or declining population, through a

1. This article is part of a longer, illustrated lecture given to Professor Casas-Torres' geography seminar at the University of Madrid in May 1973. It is based in part on Edward L. Ullman, "Geographic Prediction and Theory," in *Problems and Trends in American Geography*, ed. Saul B. Cohen (New York: Basic Books, 1967), pp. 124–45. Some corrections have been made of editorial and typographical errors, the section in the original (about six pages) on "Recent Stages in Geographic Thought" has been omitted, about nine new pages added on "Character and Connections of the Region as an Urban Field," "Travel Patterns of Fishermen Compared to Other Recreation Users," two new paragraphs at or near the end on "National Policy Implications," and other additions and changes, although the conclusions remain the same.

2. A three-volume report plus appendices resulted from the study: Edward L. Ullman, Ronald Boyce, and Donald J. Volk, *The Meramec Basin: Water and Economic Development*, Washington University Press, St. Louis, and Meramec Basin Corporation, Kirkwood, Missouri, 1962. Blair Bower, particularly, and others also contributed to the report.

to be the key catalyst of change, but at that time there was no easy or acceptable way to measure their benefits. Ullman provided the way.

Donald Volk, Blair Bower, and I worked closely with Ullman on this project. Volk, Ullman's research assistant, gathered the data, and suggested some of the analytical techniques for interpreting them. The analog procedure as presented in the article here was Ullman's creation and is included to demonstrate the application of his abstract ideas to practical purpose.

It should be said that Ullman was not truly interested in measuring recreation benefits per se, but in applying the spatial analog technique to a real area. (See my comments in A Man for All Regions, *pp. 92–107.) Nonetheless, the details required for such spatial transfer in the then murky waters of recreation benefits of a reservoir may have tended to obscure the main message of the presentation. The version reprinted here, while it lacks some details, is, I believe, the most general and philosophical discussion of the topic. This paper was mimeographed for the* Proceedings, Water Resources Conference, Department of Civil Engineering, University of Washington, 1964, *and was subsequently issued as Reprint No. 5, Center for Urban and Regional Studies, University of Washington, 1964.*

3. Many maps based on quantitative data and original surveys were prepared to delineate the character and connections of this new type region; they were published in the complete report, and used in the lecture.

belt of rapid population increase with few old people, extending out and beyond exurbia, especially along the principal axis of the region—the four lane, transcontinental superhighway, U.S. Interstate 44 (formerly U.S. 66), running along the broad ridge between the Meramec and Bourbeuse Rivers. Back from the highway and suburbs, population in most of the rural townships is declining and is characterized by a high proportion of elderly, about 15 percent over age sixty-five in 1960, compared to about 7 percent in the suburbs. Many of the young have left the farms or outer villages. Median family income also declines to only one third in the outer counties as compared to St. Louis County or city. Public tax expenditure per pupil in schools also is only about one third that of St. Louis County or city. On the southern margin of the region there is even some "open range"—unfenced grazing land—declining rapidly in extent, and found elsewhere only in the Far West of the United States today.

The whole area, however, especially that nearer St. Louis, is being brought increasingly closer to the city. Most of the region is now within two hours driving time because of the improved super highway. A mapping of telephone calls reveals a gradient from four calls per month per instrument to the central city, St. Louis, from just a few miles beyond the built-up area of the suburbs (a frequency which the Census Bureau has suggested as the limit of metropolitan area characteristics), down to one half call on the edge of the basin, just beyond the two-hour driving time limit. These outer borders in turn are closer to the city than the conventional wholesale and other trade area boundaries, and are even slightly closer in than the metropolitan newspaper circulation areas.[3] In a sense a newly recognized and newly important type of region has come into being in response to the automobile, improved highways, and increased affluence. From most of the region there is some commuting to the central metropolis; many weekend homes and cottages and much land are owned by city residents or workers, and a few branch factories for shoes and other goods or other facilities have been built. At places in the outer edge of the built-up area of the city and suburbs, a ring has developed of dispersed and clustered factories, warehouses, employment nodes, and shopping centers, all more quickly accessible to the country than the former core of the city. This is the new focus of the region for many activities, a region intermediate between the conventional metropolitan area and the larger economic hinterland.

This new type region, with typically blurred boundaries, has no established name, but the term "urban field" as used by John Friedmann and Brian Berry describes it. A precise designation in the Meramec might be "the one and one-half hour driving time," or "one and one-half telephone call per month" type region! By proposing a reservoir for recreation in such a location, we were, in a sense, taking advantage of the tension, not of the old type between city and country, but rather of a different sort between a new, close-in region, poised for development, and the old city. We sought to recognize a growing life style and to plan for a more sensible and esthetic development preserving and enhancing environmental amenities—all within an institutional framework of benefit-cost analysis, applicable to American water resource projects. Such development not only would benefit the urban residents, but also should help to improve economic conditions in an area needing additional employment opportunities.

Probably few cities are farther from desirable water recreation than St. Louis. Therefore would the rapid increase in demand for recreation and changes in technology now warrant, in a sense, building an "ocean" nearby? In other words the market should be considered a feature of increasing geographical importance in the geography of the twentieth century. Existing water recreation use already was heavy, but far away; it is shown in figure 10.1, indicating visitors at the eight large reservoirs in the St. Louis universe, with pie-shaped sectors showing estimated St. Louis visitor-days.

THE PROBLEM: PREDICTING ATTENDANCE
AND BENEFITS FROM WATER RECREATION

The problem was to try to estimate how much benefit would result from providing a large lake for recreation. The costs could be readily figured out, but estimating the benefits was an unsolved problem. Benefits of course should exceed the costs. This is the classic way in which American water resource problems are investigated, and indeed, the same procedure is being tried on other problems.

Recreation, as noted, however, is an intangible item, difficult to measure. In fact, economists had formerly thrown up their hands at the problem but we, rushing in where angels feared to tread, did invent a method for measuring the benefits of recreation. This method although developed independently, turned out to be related to suggestions and work of Hotelling, Clawson, and others as noted later in this treatment.

"Finally, the invention comes out of a whole facet of previous geographic and related thinking on other topics ranging from von Thünen's Isolierte Staat of 1827 through Walter Christaller's Central Place Theory of 1935. Even more specifically related is William Applebaum's pioneer and subsequent work on store location started in the 1930's, which has had great practical utility. In recent work Applebaum also uses the term analog to describe his method of plotting per capita sales by bands around stores, similar to the method followed in Table 1, William Applebaum, 'Methods for Determining Store Trade Areas, Market Penetration, and Potential Sales,' *Journal of Marketing Research*, Vol. III (May 1966, pp. 127–41.)

My work on this is one of the most important things I have done, although not widely known. It is somewhat similar to the work of Clawson, Hotelling, and Knetsch." [Letter to Charles E. Cuningham, 24 August 1972]

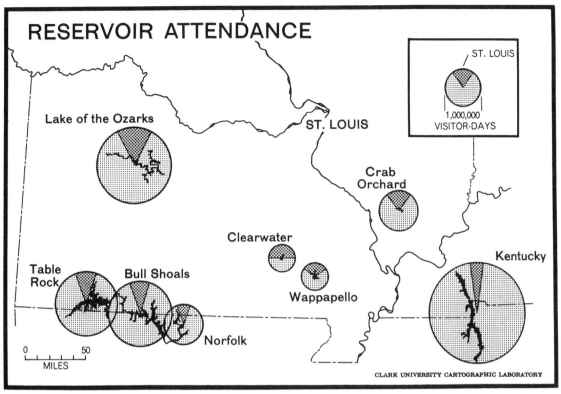

Fig. 10.1

The method invented illustrated what I would call *geo-graphical prediction*. The term is employed very precisely here. Normally, one might say there is only one kind of prediction—simply the projection of a trend such as population increasing so much per decade, and projecting this trend into the future. Any other kind of prediction often involves guessing, using a crystal ball or some unscientific approach. Thus, most valid prediction is simply an adjusted projection of an existing trend. This might be termed *historical* prediction.

But what does one do when there is no trend to project—when building something new, for example? We need to know (1) how many visitors would attend a new lake, and (2) what the number of visits was worth.

This calls for inventing a second method of prediction, as noted, which I shall call *geographical prediction*. This means that if one does not have a trend to project in one area, one looks for essentially analogous conditions elsewhere, transfers that setting to the study area, making appropriate adjustments, and uses this for prediction. This could be called the *geographical analog* method. It is a natural by-product of comparative geography and can be used widely; it is also related to cross-sectional analysis in economics. I first developed the idea as a device for predicting traffic on a proposed railroad line in the Philippines in 1956.

In the operation of this particular model of prediction, advantage is also taken of the quantitative approach (so characteristic of geography today—1973) by ordering the data into either a gravity model or simply a distance regression line. An interaction or functional approach is also employed. We are interested in where people have gone. Our fundamental theory is that number of visitors declines somewhat in proportion to distance—a version of the gravity model.

STEP 1: ESTIMATING ATTENDANCE

In implementing the model, Donald J. Volk was a co-worker with me, and gathered and processed most of the data which follow. Data were obtained from the Missouri Conservation Commission on the origin of fishermen who had caught fish at many lakes in the State of Missouri. Altogether there were one hundred thousand or more of these origins. These origins were fugitive data acquired merely as by-products of catching fish. Oftentimes more knowledge is available for fish than for people! Fortunately, we found these data, but one does not find data unless one knows what to look for—in other words, has a theory. If we had not found these data, we would have extended our own surveys to provide substitute data.

Fishermen as a Proxy for All Visitor Travel

Fishermen's travel is close to the average of most other recreation travel, especially at the fifty-mile (eighty-kilometer) range, the distance of most concern to the study. Subsequent surveys prepared by the project and conducted by the Missouri State Parks Commission at four lakes established this (fig. 10.2). Thus respondents answering questions given to a stratified sample of visitors indicated that at fifty-mile distance from their homes 36 percent expected to engage in fishing, 37 percent in relaxing, and 30 percent in motor boating. Balancing these at the upper end of the spectrum were about 48 percent swimming, 49 percent sightseeing, and

"In the beginning I noticed you do not mention my enclosed study on the Meramec, [for] which I don't blame you because it was more specifically oriented to attendance and benefits. However, that study was by far the most complete and I think the best of those applying the travel savings version of the consumer's surplus model. In fact Trice and Wood are wrong in that they assume that everybody gets the benefit of the 90 percentile traveler. Likewise, your statement on page 10 'the difference between the price paid by local consumers and those from farthest away provide the measure of the financial recreation benefit,' is apparently what Hotelling and Trice and Wood are saying, but this is not correct. What is needed is a measure of how far people will travel and then seeing how much additional they will get if they don't travel this far, as I explain in the enclosed article. In other words it is not the most distant traveler or even the 90 percentile. Hotelling in his famous letter, if I understand it, made the same mistake." [Letter to W. R. D. Sewell, 29 December 1970]

57 percent picnicking; and at the lower end, 23 percent water skiing, 17 percent camping, and 4 percent other boating. Most respondents indicated planned participation in several (generally at least three) recreation activities.

For our purposes somewhat the same pattern emerges when respondents were asked to list their *one* most important activity. Results from Table Rock and Lake of the Ozarks are probably the most meaningful, since they are the larger reservoirs most comparable to a proposed Meramec one. At Table Rock, 32 percent listed fishing, 6 percent relaxing, 3 percent motor boating, the activities which were bunched at the fifty-mile (eighty-kilometer) distance on the participation chart (fig. 10.2). Total of these is 41 percent. In addition 27 percent indicated that all activities would be about equal, reaching a total of 68 percent. For Lake of the Ozarks, 13 percent listed fishing as the most important single use; 13 percent, relaxing; 17 percent, motor boating, for a total of 43 precent, and 15 percent rated all about equal, achieving the same total as Table Rock of 68 percent. At both lakes combined, less than 5 percent listed both picnicking and other boating which, in percent participating at fifty miles (eighty-kilometers), were the activities diverging the greatest from fishing (fig. 10.2).

The precise behavior of fishermen in any one year, of course, is unpredictable. In some years some lakes put out more fish, in others less, which affects fishermen's destinations. Thus Table Rock, somewhat like most newer reservoirs, has a deserved, high reputation for fishing. Furthermore, the results may be slightly biased there, since the survey location was near the fishing docks and swimming beaches had yet to be developed, while conversely Lake of the Ozarks State Park had good swimming beaches and the survey location was close to camp grounds. If fishing is less important, however, other uses will rise to compensate to some extent. Since fishing is a good proxy for the average of all uses at fifty miles (eighty kilometers) in Missouri, it can still serve as a prediction agent for the average situation over a long time period, as useful as any other measure which can reliably be derived from the data.[4]

Visitor Attendance by Distance

The origins of fishermen were plotted by counties and converted to per capita measures for the five-year period on maps. The number, in general, declined with distance (fig.

4. The explanation in the three paragraphs above is longer than in the original short paper in order to respond to previous readers' desire to have more information on this point. In the survey about 2,000 cars were questioned containing an average of about 3.8 persons per car, with about 240 cars at Table Rock and 450 at Lake of the Ozarks. The participation question read: "Which of these things do you plan to do at this lake (or park)?" The second question was: "Is any one of these things more important than the others as a reason for your coming here? If yes, which one?" For further details see "1960 Missouri State Park Recreation Survey," appendix to chap. 5, vol. 3, *The Meramec Basin* (St. Louis: Washington University, December 1961). See for example the review of *Problems and Trends in American Geography*, by A. G. Isachenko in *Soviet Geography* 11 (1970):678–81, which in turn was translated from *Izvestiya Vsesoyuznogo Geograficheskogo Obshchestva*, 1970, no. 1, pp. 96–98 and my reply to this and other points in *Soviet Geography* 12(1971):240–41.

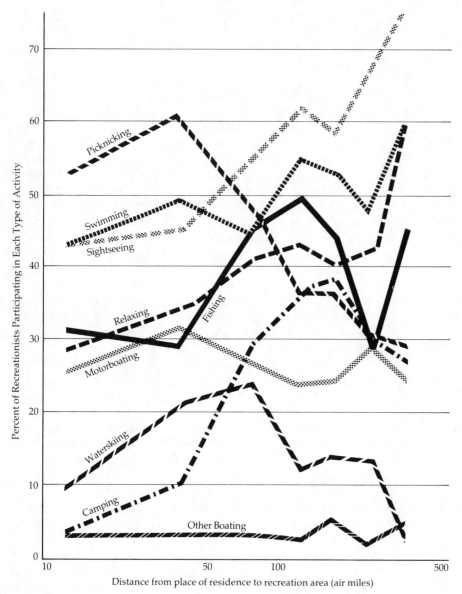

Fig. 10.2

10.3; this is one of numerous maps for various reservoirs in the original report). These data were also plotted on a scatter diagram with logarithmic scales; the decline in attendance was proportionate to a range varying from the square of the distance up to the distance to the fourth power (fig. 10.4).

Lake of the Ozarks, Niangua Arm
INTENSITY OF USE BY FISHERMEN

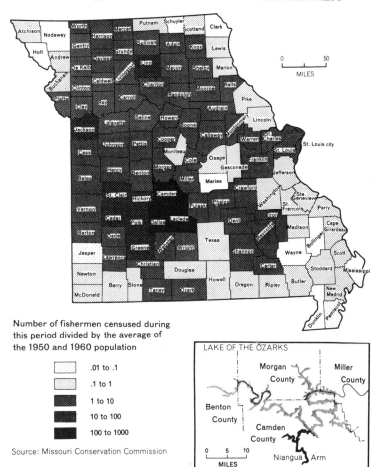

Number of fishermen censused during
this period divided by the average of
the 1950 and 1960 population

	.01 to .1
	.1 to 1
	1 to 10
	10 to 100
	100 to 1000

Source: Missouri Conservation Commission

LAKE OF THE OZARKS

CLARK UNIVERSITY CARTOGRAPHIC LABORATORY

Fig. 10.3

On the average, the decline in attendance was proportionate to distance almost to the third power, indicating a rather rapid decline in attendance with distance. This indicates further probably that one reservoir or lake is more or less substitutable for another, and therefore the fishermen went to the nearer-by impoundments. This apparently contrasts to some of the more nearly unique natural features of the West, such as Yellowstone Park or Grand Canyon where decline in attendance appears to be more gradual, resembling distance squared.

On the average we found also that we tended to get somewhat more visits per capita from urban or high income counties than expected, and less from lower income counties or
from counties in which there were intervening opportunities—that is, another lake in-between. Three lines were
therefore plotted on the graph and converted to annual
figures, a median line, an urban high income line, and a rural,
low income line; this became our prediction model (fig. 10.5).
From this, with some adjustments, we could predict annual
attendance anywhere we put down a lake. We could read off
a certain distance on the horizontal axis and find, for example, at fifty miles the visits from an urban high income county
would be four visits per capita per year.

For nearby distances, especially less than about twenty-five
miles (forty kilometers), our prediction model overestimates
attendance somewhat, which may be a reflection of Roy
Wolfe's later "Start-up" or inertia model, based on Ontario,
Canada, recreational surveys: visitation appears to be less
than expected at short distances because of the effort to start a
trip.[5] However the use of a high distance decline rate, close to
the third power, and its application primarily to the fifty-mile
(eighty-kilometer) range probably makes any adjustment unnecessary.

This model was operated for a variety of locations around
St. Louis, and tables made for each location simply by drawing circles of twenty-mile bands around each impoundment
(table 10.1), multiplying the population in the circular zone by
the per capita attendance expected for that distance, to
produce an estimate of the number of visitor-days per year.
The biggest contributor in all these was the St. Louis metropolitan area with its two million people. For one hypothetical lake fifty miles from St. Louis (table 10.1) this came out to
eight million visitor-days per year from the St. Louis metropolitan area alone, which would have produced the highest
attendance at any reservoir in America. This was quite logical
because this reservoir would have been only fifty miles from a
large metropolitan area.

The prediction in this case was based on the attendance
which resulted from building a reservoir with the average
characteristics of the large ones in our analog origins (fig.
10.1). As a check on our prediction model, we plotted attendance from a variety of cities in the country to nearby reservoirs for as many as we could get (fig. 10.5). The plots all fell
on the lines of the graph more or less as they should, save
perhaps some at short distances, as already noted. As a matter of fact, St. Louis visitor-days to five large impoundments

"I am used to others (economists especially) not paying attention to the report, and
hence am at least grateful to Mr.
Isachenko for noticing it. Perhaps people do not want an apparently simple demonstration
of the utility of geography. In
point of fact it was not at all
simple—quantities of original
back-up data were gathered and
processed in the course of two
years' work . . . by using an
analog approach which permitted retention of all the variables
in a total system including the
physical ones. . . . We used a
real piece of earth to make a real
prediction for a real place. This
is an example of pushing geographical science and art to its
ultimate as a legitimate, a comparative, and an original contribution." ["Reply to a Review
by A. G. Isachenko," *Soviet
Geography* 12 (1971):240–41]

5. Roy I. Wolfe, "Distance of Vaca-
tion Homes, Environmental Preferences, and Spatial Behavior," *Journal
of Leisure Research* 2 (Winter 1970):
85–87, plus other contributions on
"Start-up" theory.

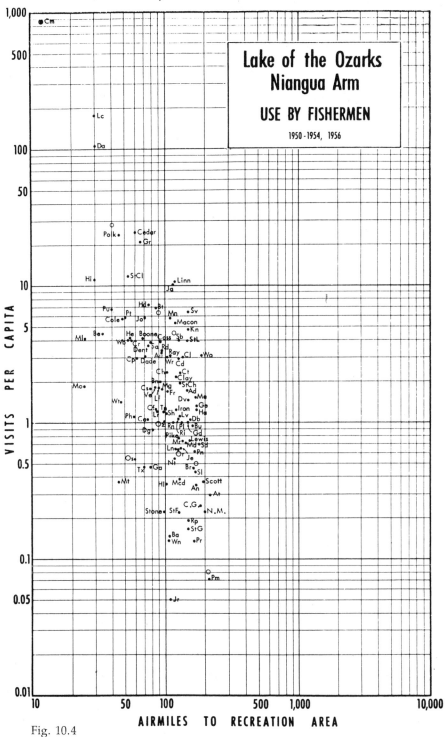

Fig. 10.4

TABLE 10.1

ESTIMATED ATTENDANCE AT HYPOTHETICAL MAJOR IMPOUNDMENT
50 AIR LINE (60 ROAD) MILES FROM ST. LOUIS

Distance Zones from Hypothetical Reservoir	Population in Zone	Per Capita Category[a]	Annual Visits Number	Total Annual Visitor Days
0–10	20,000	median	15+	300,000
10–20	40,000	median	10+	400,000
20–40	90,000	high	10+	900,000
40–60	2,000,000[b]	high	4	8,000,000
60–100	300,000	low	0.4	120,000
100–200	2,000,000	median	0.03	60,000
200–300	10,000,000	median	0.005	50,000
300–500	30,000,000	median	0.001	30,000
Beyond 500	(Rest of U.S.)		Estimate	50,000
			Total	9,910,000

[a]Categories refer to per capita annual visits expected based on numerous origin surveys in 1960, at large impoundments in Missouri, Illinois, and Kentucky. *High* refers to per capita expectation from urban, high income and/or lack of intervening opportunities for recreation at nearer impoundments; *low* refers to rural, low income and/or intervening opportunities for recreation at closer impoundments. *Median* is average between two extremes.

[b]Includes most of St. Louis metropolitan area.

from 100 to 125 miles away were all in the high zone in per capita attendance, whereas St. Louis visitor-days at three other smaller impoundments were in the median zone. Without exception those in the high zone were the better and larger impoundments and those in the median zone were poorer and smaller impoundments. The variation was in the expected direction. Thus we feel that the prediction model is reasonably valid.

Although we acquired more meaningful quantitative data on origins and characteristics than any previous study, certain inherent defects of the data argued against still more precise calculations, including fitting of a least-squares regression line and calculating some coefficient of correlation of the scatter of observations, as on figure 10.4. Figure 10.4, however, represents only the most important of the several lakes and surveys used as a base for the composite prediction model, figure 10.5. Even more important, as noted, was the use of an independent check from city attendance at other reservoirs to confirm our prediction at about the 25–150 mile

"P.S. I wonder whether the general method used by Applebaum and me, even including the use of the word 'analog,' might be taken as a good example of geographic thinking to the solution of problems. Neither of these approaches has received too much attention nor have they been explicitly written up very much in the geographic literature, perhaps because we have only just now finalized them." [Letter to Edward Taaffe, 31 October 1967]

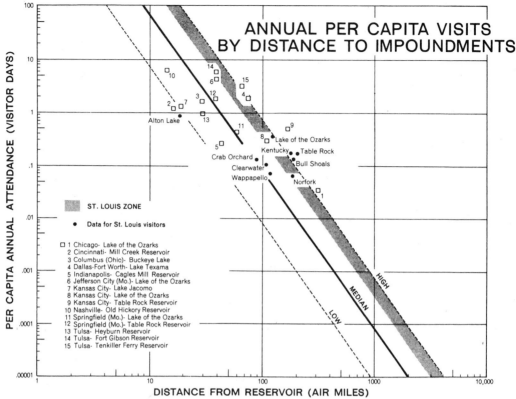

Fig. 10.5

(40–240 kilometer) range. However, it might well be desirable to include more precise regression and correlation measures, after calculating separately the slope for urban, high income counties, as opposed to low income, rural and/or intervening opportunity counties. We have already observed that the high income counties, generally, but not always, have higher per capita attendance (i.e., tend to be on the right hand side of a visually drawn trend line on figure 10.4). Conversely the low income counties tend to have lower attendance, but more scatter. Perhaps the measure should be based on a certain percentage or number of families with a stated higher income or other characteristics; this would pick up rural, low income counties containing a small city or town of higher income. In making these finer distinctions one faces the danger of running out of counties with a large enough attendance to avoid sampling errors due to small number of visitors. To overcome this we used records as long in years as we could obtain. However, perhaps something more could be done by combining more cases. The problem, however, as is so often the case in original geographic research, is the data, not the manipulation, although the two go together.[6] Spurious precision is no virtue.

STEP 2: ESTIMATING BENEFITS

In addition to attendance, we needed to know what these visitor-days were worth. This was estimated in two ways, first, by a primitive diversion model and second, by a generated and diverted visitor model.

The Diversion Model

As a first approximation, a diversion model was devised, based on how many St. Louis visitor-days might be diverted from the eight impoundments in the St. Louis universe (fig. 10.1) lakes ranging from 100 to 125 miles away, by building a lake only 50 miles away.[7] The diversion was estimated on the basis of surveys made at various Missouri State Parks based on the question: "If a lake similar to this one were built half as far away from your home, would this decrease your visits to this lake to the extent of eliminating completely, reduce greatly, reduce slightly, no effect, don't know." These were arbitrarily scored 100 percent, 75 percent, 25 percent, zero, and the last, don't know, responded by about 20 percent, allocated proportionately. The combined score of this rough and ready approximation indicated a diversion of about one-third of the existing visitors if a lake were built only fifty or sixty miles from St. Louis. This was a conservative figure; at first we thought this "a fortiore" approach would give more than enough benefits to justify construction. After working out some of the estimates, however, we found that the diversions alone did not give quite enough benefits, although the operation of this model gave somewhat of an independent check on the subsequent model.

Generated and Diverted Visitors: A Travel Savings Model

A more complete measure was needed to give the value of newly generated business as well as diverted business. How would one measure this generated business? This we were able to work out quite simply, in retrospect, from our attendance prediction model itself (figure 10.5).

The rationale is as follows: if a reservoir were built fifty miles from St. Louis, according to the graph (fig. 10.3) eight million visitor-days would be willing to go fifty miles or

6. Estimates of total attendance at reservoirs, although checked in various ways, also vary. Not the least of the problems is the definition of a "visitor day" which can range from a ten minute stop to view a dam, to an all day picnic or fishing expedition. For this reason, in many of our actual calculations in the Meramec study, we arbitrarily cut the visitor day estimates in half and made other adjustments. For the purposes of table 1, however, we have not done this so that the figures can be compared nationally. For planning purposes we subsequently made extrapolations over the usual fifty-year period, discounted at stated percents, and for a four- or five-fold increase, reflecting increased affluence and population as estimated by O.R.R.R.C. (For details see vol. 1 of the Meramec report.)

7. For detailed figures see the Meramec report, or Edward L. Ullman and Donald J. Volk, "An Operational Model for Predicting Reservoir Attendance and Benefits," *Papers of the Michigan Academy of Science, Arts and Letters* 47(1962):473–84.

farther, but 4,600,000 of those would be willing to go sixty miles, 2,800,000 would go seventy miles, and so on (table 10.2). These first 3,400,000 visitors (the difference between eight million and 4,600,000) are in essence being given a gift of ten miles. They are willing to pay to go sixty miles. But, if a new reservoir is provided only fifty miles away, each of them is given a gift of ten miles, or a twenty-mile round trip, etc.

The next question was to figure out what these saved miles were worth. We did this by putting a value on them of about three cents per visitor-day-mile, comprised half of modest estimates of vehicle out-of-pocket or variable operating costs—gasoline, oil, tires, etc., but not depreciation. The other half consisted of estimates of the value of time which

TABLE 10.2
POTENTIAL TRAVEL SAVINGS FOR IMPOUNDMENT
50 MILES FROM ST. LOUIS

Distance from St. Louis (Air miles)	Visitor (Days) Willing to Go Distance in Col. 1, or Farther	Approximate Visitor (Days) in Each Incremental Ten-Mile Block	Travel and Time (cents per visitor-day[a])	Cost Saving Approximate Total (Col. 3×4)
1	2	3	4	5
50	8,000,000			
60	4,600,000	3,400,000	9	$304,000
70	2,880,000	1,700,000	39	669,000
80	1,920,000	960,000	69	658,000
90	1,340,000	580,000	99	575,000
100	980,000	360,000	129	470,000
110	734,000	246,000	159	390,000
120	564,000	170,000	189	321,000
130	443,000	121,000	219	265,000
140	354,000	89,000	249	222,000
150	287,000	67,000	279	187,000
160	236,000	51,000	309	158,000
170	197,000	39,000	339	132,000
180	165,000	32,000	369	118,000
190	140,000	25,000	399	100,000
200	120,000	20,000	429	86,000
210	104,000	16,000	459	73,000
			Total	$4,728,000
			($.59 per visitor-day)	

[a] At $.03 per visitor-day mile × one-half round trip distance (represents round trips divided by two days, the average duration of trip as determined by survey at Meramec State Park).

were taken from standard American highway manuals and averaged with other estimates we had made.[8]

Economists are often unhappy with estimates of the value of time. They query: How do you know what the value of time is? My first answer is, "I don't care." If we are wrong, by say 50 percent, it still affects our total figure by only about 25 percent plus or minus, because half of those dollar savings were simply the operating costs of the automobile. Personally, I think the value of time, on the average, is more important in choosing a destination than the value of gasoline and oil used on the trip.

This whole model applies to America where the automobile is overwhelmingly the vehicle used for this type of recreation. In other parts of the world, where buses and other vehicles are also used, this would not be so true; but similar measures could be readily, in fact more easily, made, by using actual fares paid.

Each ten-mile block (columns 1, 2 and 3 in table 10.2) is multiplied by the unit savings (col. 4) based on the following logarithmic block averages: three miles, thirteen, twenty-three, thirty-three, etc., and the savings are totaled in the last column (col. 5). A total savings of $4,728,000 results from the blocks of visitors willing to go to an impoundment fifty miles from home. Next the eight million visitor-days (the theoretical total unadjusted attendance) was divided into $4,728,000 to produce fifty-nine cents per visitor-day, or rounded off to sixty cents. This was the actual figure used in our visitor-day calculations—sixty cents benefit per visitor-day.

A value per visitor-day is a convention commonly used; we are not too happy with this convention, but the $4,728,000 can be used any way one wants. This is a *net national* saving. It has the virtue of not having any double counting or other tricks in it.

This particular model, which I might call a "geographical analog, spatial interaction model" if I were to use geographic language, is similar to the "travel savings version of a consumer's surplus model" from economics. Here geographic thinking and economic thinking meet in a common end.

The first to suggest this method, in a preliminary way, was Harold Hotelling, Professor of Statistics at the University of North Carolina, who apparently argued for giving everyone, regardless of distance traveled, as benefits, the cost incurred by the longest distance traveled.[9]

Trice and Wood actually operated a somewhat similar method in California, but based it on a very small sample, to

8. Based on variable vehicle operating costs of 5.3 cents per vehicle mile (as reported by Wilbur Smith and Associates, *Future Highways and Urban Growth*, New Haven, 1961, p. 281) divided by 3.5 passengers per car, as determined from Meramec Basin-Missouri State Parks survey of our universe, equals 1.5 cents, plus time savings valued at $1.35 per hour per passenger car (ibid., p. 285), divided by 3.5 passengers per car equals 38 cents per vehicle hour divided by 40 miles per hour average speed equals about 1 cent to produce a total of about 2.5 cents per visitor-day mile. On the other hand, estimating 85 cents per visitor-day hour (based on $2.00 per hour for driver, $1.00 for 1 passenger and nothing for the other 1½ as indicated in table II, IIC of Ullman & Volk, "Predicting Reservoir Attendance") results in more than 2 cents per mile or more than 3.5 cents total savings per mile. Average is taken as approximately 3 cents per mile, therefore.

9. From personal conversation I believe he would have modified this, if he had developed the idea further, essentially along the lines used in the Meramec study but with different methods. His suggestion was contained in a letter to the National Park Service shortly after World War II.

arrive at a figure of $2.00 per visitor day. This figure was based on the difference between the median distance traveled and almost the longest distance, the ninetieth percentile. This cut down the amount presumably suggested by Hotelling, but presents no persuasive rationale for taking even this great a benefit.[10] Subsequently, Marion Clawson developed the hypothesis systematically in 1959 and plugged in a few suggestive attendance data at National Parks.[11] Subsequent work by J. L. Kentsch, Roy I. Wolfe, and others carried the methods further.

Optimum Admission Fee

One other way to use the data was to imagine that a fence would be built around the reservoir, a gate constructed, and admission charged—a classic economic solution. What admission fee would give the greatest return? This we were able to calculate (fig. 10.6). We found that about one dollar would give the greatest gross return, $1.29 the greatest net return, if one postulates a reasonable twenty cents per visitor-day maintenance cost. (Total visitors at Missouri State parks divided by the annual expenditure was eleven cents per visitor-day in 1960.) There would be fewer visitors with a fee, of course, than if nothing were charged. This does not mean that this is the exact amount that should be charged, as Marion Clawson has noted, but rather that, since the top of the revenue curve is rather flat, a range, in this case from about 50 cents to $2.00 would yield close to the greatest revenue. This knowledge should be useful to policy makers facing a decision. Thus, we also solved this economic problem by use of a spatial or geographical approach.

NATIONAL POLICY IMPLICATIONS

This study, in addition to strengthening recreation as a purpose in subsequent Corps of Engineers' proposals for development of the region, also has influenced national policy. During the early stages the findings were reported to the Outdoor Recreation Resources Review Commission; the importance demonstrated by the findings of locating a reservoir close to a large center influenced what was perhaps the most important recommendation of the Commission: to develop outdoor recreation near large urban centers.[12] Subsequent policy statements by the Recreation Advisory Council and the new Bureau of Outdoor Recreation of the Department of Interior embody this recommendation as, for example: "Primary

10. Andrew H. Trice and Samuel E. Wood, "Measurement of Recreation Benefits," *Land Economics* 34 (1958):195–207.

11. M. Clawson, *Methods of Measuring the Demand for and Value of Outdoor Recreation* (reprint no. 10, Resources for the Future, Inc., 1959). J. L. Knetsch, "Outdoor Recreation Demands and Benefits," *Land Economics*, November 1963; "The Influence of Reservoir Projects on Land Values," *Journal of Farm Economics*, February 1964, p. 231; "Economics of Including Recreation as a Purpose of Water Resources Projects," to be published in the 1965 *Proceedings* of the American Farm Economics Association. For a general assessment of outdoor recreation research possibilities, see M. Clawson and J. L. Knetsch, *Outdoor Recreation Research* (reprint no. 43, Resources for the Future, Inc., 1963). Since then many of these studies have been further assessed in Marion Clawson and Jack L. Knetsch, *Economics of Outdoor Recreation* (Resources for the Future, Inc., 1966). The 1971 reprint of the book is carefully reviewed by R. I. Wolfe, in *The Canadian Geographer* 16 (1972):195–96, who notes that in the type of economic analysis used, little has changed. See also Robert J. Smith, "The Evaluation of Recreation Benefits: The Clawson Method in Practice" (University of Birmingham, 1972) and the broader survey: Francoise Cribier, "La géographie de la récréation Anglo-Saxonne," *Annales de géographie* (1972):644–65.

12. Outdoor Recreation Resources Review Commission, *Outdoor Recreation for America*, Washington, D.C. 1962.

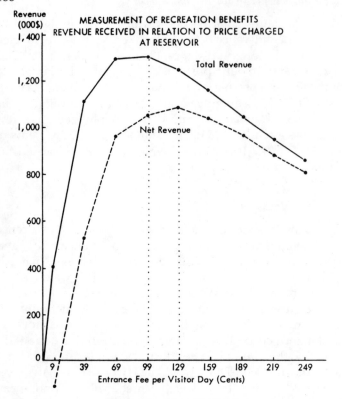

MEASUREMENT OF RECREATION BENEFITS
REVENUE RECEIVED IN RELATION TO PRICE CHARGED
AT RESERVOIR

	Annual Revenue (Dollars)		
Entrance Fee (Cents)	Number of St. Louis Visitor Days	Total	Net (Assuming 20 Cents per Visitor Day Maintenance)
0	8,000,000	0	−1,600,000
9	4,600,000	414,000	−106,000
39	2,880,000	1,123,200	547,000
69	1,920,000	1,324,800	940,800
99	1,340,000	1,326,660	1,058,600
129	980,000	1,264,200	1,088,200
159	734,000	1,167,060	1,021,000
189	564,000	1,064,210	951,460
219	443,000	970,170	881,570
249	354,000	881,460	800,660

Add c. 20 per cent for all visitors

Fig. 10.6

priority for outdoor recreation investment normally shall be assigned to projects which are: (a) located near (within approximately 3 hours automobile travel distance) urbanized areas according to the most recent decennial census; and/or (b) to be acquired and developed in conjunction with water bodies (reservoirs, natural lakes, seashores and rivers)."[13]

13. Recreation Advisory Council (Secretaries of Agriculture, Commerce, Defense, HEW, Interior and HUD), *Guides for Federal Outdoor Recreation Investment*, circular no. 5, Washington, D.C., 1965, p. 2. See also the fifth primary criterion for selection of national recreation areas: "Although non-urban in character, National Recreation Areas should nevertheless be strategically located within easy driving distance, i.e., not more than 250 miles from urban population centers which are to be served." The first of the specific secondary criteria also gives preference to an area to be located: "Within or closely proximate to those official U.S. Census Divisions having the highest population densities" (*Federal Executive Branch Policy Governing the Selection, Establishment and Administration of National Recreation Areas*, Washington, D.C., 26 March 1963, circular 1). See also an extension of this line of thinking: "In California, the definition of purpose for the Outdoor Recreation program has been extended to emphasize projects which are '. . . regional in significance, and primarily serve day use from urban centers of population, preferably within approximately one hour's auto travel distance from such centers'" in *State of California, Outdoor Recreation Local Grant Program Manual*, 22 September 1965, p. 6; reprinted in *Recommended Roles for California State Government in Federal Urban Programs*, a report of the Intergovernmental Council on Urban Growth, Sacramento, 1967, p. 30.

The Recreation Advisory Council statement was also used to advantage in arguing for turning Fort Lawton in Seattle into a park. (Edward L. Ullman, letter, *Seattle Post-Intelligencer*, 17 December 1968, and "The Urban Problem and the University," *University of Washington Report*, vol. 1, no. 2, 27 April 1970, p. 7.)

CONCLUSION: WHAT CREATES AN IDEA?

This case of geographical prediction is presented because it does represent a breakthrough—an invention, if you please, which I wanted to share with you. This brings up the question of the value of an applied job. Oftentimes an applied job is considered less desirable than a theoretical or academic approach. But many times the reverse can be true. It is true that "leisure to experiment and reflect" is necessary as Carl Sauer notes in speculating about the development of human learning.[14] If one is forced to come up with an answer, however, one's head, so to speak, is pushed up against the wall. Sometimes an imaginary answer is given; sometimes a real answer can be invented.

This in turn brings up the question as to what does produce a creative idea? I by no means know what fundamentally gives one an idea, but two or three conditions seem to be helpful: (1) *interaction* with somebody with a slightly different way of thinking or approach, and (2) *necessity*. This is the old saw of necessity being the mother of invention. Both these factors help to give one an original idea and to make an original contribution, especially if enough time is also available to reflect and experiment. A gestation period appears to be necessary to develop the idea, followed by secondary breakthroughs, somewhat analogous to the stages in the firing of a rocket.

Finally, the invention emerges from the background of previous geographic and related thinking on other topics ranging from von Thünen's *Isolierte Staat* of 1827 through Walter Christaller's Central Place Theory of 1935.[15] Even more specifically related is William Applebaum's pioneer and subsequent work on store location started in the 1930s, which has had great practical utility. In recent work Applebaum also uses the term *analog* to describe his method of plotting per capita sales by bands around stores, similar to the method followed in table 10.1.[16] One is tempted to cite still another cliché to the effect that there is nothing new under the sun, but instead let us say that one owes a conscious and unconscious debt to previous inventors and to the discipline of space and geography as practiced by its more thoughtful innovators.

14. Carl O. Sauer, "Environment and Culture during the Last Deglaciation," *Proceedings of the American Philosophical Society* 92 (1948):65–77. Also in Carl O. Sauer, *Land and Life* (Berkeley: University of California Press, 1963), p. 259. He notes: "The hearths of human learning needed to be somewhat sheltered from the world at large, but also to have the option of outside communication."

15. Cf. Edward L. Ullman, "A Theory of Location for Cities," *American Journal of Sociology* 46 (1941):853–64; and Brian J. L. Berry, *Geography of Market Centers and Retail Distribution* (Englewood Cliffs, N.J.: Prentice-Hall, 1967).

16. William Applebaum, "Methods for Determining Store Trade Areas, Market Penetration, and Potential Sales," *Journal of Marketing Research* 3 (1966):127–41.

PART IV | Urbanization

"Cities are the focal points for the occupation and utilization of the earth by man. Both a product of and an influence on surrounding regions, they develop in definite patterns in response to economic and social needs.

"Cities are also paradoxes. Their rapid growth and large size testify to their superiority as a technique for the exploitation of the earth, yet by their very success and consequent large size they often provide a poor local environment for man. The problem is to build the future city in such a manner that the advantages of urban concentration can be preserved for the benefit of man and the disadvantages minimized." [Chauncy D. Harris and Edward L. Ullman, "The Nature of Cities" (*Annals of the American Academy of Political and Social Science*, vol. 242, November 1945), p. 7]

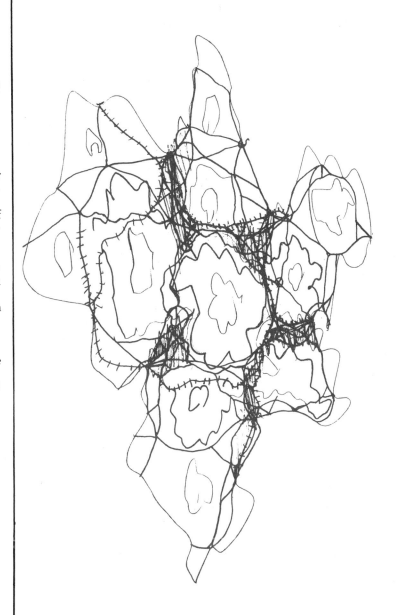

CHAPTER 11   A Theory of Location for Cities

This classic article is perhaps the pioneer work on the distribution of cities to appear in America in this century. Theories about central place were published initially by Walter Christaller in 1933 in Die zentralen Orte in Süddeutschland. *Edward Ullman introduced Christaller's work to an English-speaking audience, extended and clarified it, and coined the phrase "central place theory" in the article reprinted here. The paper also dealt with periodic service centers and the urban economic base, ideas that Ullman followed up in later articles.*

This article has an unusual history. First, it was published by a young and obscure geographer; and second, it was not published in a geographical journal, but in the American Journal of Sociology *(May 1941), pp. 853–64. Many geographers did not see the work*

I

Periodically in the past century the location and distribution of cities and settlements have been studied. Important contributions have been made by individuals in many disciplines. Partly because of the diversity and uncoordinated nature of the attack and partly because of the complexities and variables involved, a systematic theory has been slow to evolve, in contrast to the advances in the field of industrial location.[1]

The first theoretical statement of modern importance was von Thünen's *Der isolierte Staat*, initially published in 1826, wherein he postulated an entirely uniform land surface and showed that under ideal conditions a city would develop in the center of this land area and concentric rings of land use would develop around the central city. In 1841 Kohl investigated the relation between cities and the natural and cultural environment, paying particular attention to the effect of transport routes on the location of urban centers.[2] In 1894 Cooley admirably demonstrated the channelizing influence that transportation routes, particularly rail, would have on the location and development of trade centers.[3] He also called attention to break in transportation as a city-builder, just as Ratzel had earlier. In 1927 Haig sought to determine why there was such a large concentration of population and manufacturing in the largest cities.[4] Since concentration occurs where assembly of material is cheapest, all business functions, except extraction and transportation, ideally should be located in cities where transportation is least costly. Exceptions are provided by the processing of perishable goods, as in sugar centrals, and of large weight-losing commodities, as in smelters. Haig's theoretical treatment is of a different type from those just cited but should be included as an excellent example of a "concentration" study.

In 1927 Bobeck[5] showed that German geographers since 1899, following Schlüter and others, had concerned themselves largely with the internal geography of cities, with the pattern of land use and forms within the urban limits, in contrast to the problem of location and support of cities. Such preoccupation with internal urban structure has also characterized the recent work of geographers in America and

1. Cf. Tord Palander, *Beiträge zur Standortstheorie* (Uppsala, Sweden, 1935), or E. M. Hoover, Jr., *Location Theory and the Shoe and Leather Industries* (Cambridge, Mass., 1937).
2. J. G. Kohl, *Der Verkehr und die Ansiedlungen der Menschen in ihrer Abhängikeit von der Gestaltung der Erdoberfläche* (2d ed.; Leipzig, 1850).
3. C. H. Cooley, "The Theory of Transportation," *Publications of the American Economic Association* 9 (1894):1–148.
4. R. M. Haig, "Toward an Understanding of the Metropolis: Some Speculations Regarding the Economic Basis of Urban Concentration," *Quarterly Journal of Economics* 40 (1926):179–208.
5. Hans Bobeck, "Grundfragen der Stadt Geographie," *Geographischer Anzeiger* 28 (1927):213–24.

other countries. Bobeck insisted with reason that such studies, valuable though they were, constituted only half the field of urban geography and that there remained unanswered the fundamental geographical question: "What are the causes for the existence, present size, and character of a city?" Since the publication of this article, a number of urban studies in Germany and some in other countries have dealt with such questions as the relations between city and country.[6]

II

A theoretical framework for study of the distribution of settlements is provided by the work of Walter Christaller.[7] The essence of the theory is that a certain amount of productive land supports an urban center. The center exists because essential services must be performed for the surrounding land. Thus, the primary factor explaining Chicago is the productivity of the Middle West; location at the southern end of Lake Michigan is a secondary factor. If there were no Lake Michigan, the urban population of the Middle West would in all probability be just as large as it is now. Ideally, the city should be in the center of a productive area.[8] The similarity of this concept to von Thünen's original proposition is evident.

Apparently many scholars have approached the scheme in their thinking.[9] Bobeck claims he presented the rudiments of such an explanation in 1927. The work of a number of American rural sociologists shows appreciation for some of Christaller's preliminary assumptions, even though done before or without knowledge of Christaller's work and performed with a different end in view. Galpin's epochal study of trade areas in Walworth County, Wisconsin, published in 1915, was the first contribution. Since then important studies bearing on the problem have been made by others.[10] These studies are confined primarily to smaller trade centers but give a wealth of information on distribution of settlements which independently substantiates many of Christaller's basic premises.

As a working hypothesis one assumes that normally the larger the city, the larger its tributary area. Thus there should be cities of varying size ranging from a small hamlet performing a few simple functions, such as providing a limited shopping and market center for a small contiguous area, up to a large city with a large tributary area composed of the service areas of many smaller towns and providing more complex services, such as wholesaling, large-scale banking, specialized retailing, and the like. Services performed purely

until ten years later when it appeared in Reader in Urban Sociology, *edited by P. K. Hatt and A. J. Reiss, Jr. (1951). The first geography publication that carried the paper was* Readings in Urban Geography, *edited by Harold Mayer and Clyde Kohn in 1959, eighteen years after its initial appearance. Since the publication in 1966 of the first full English translation of Christaller's study by Carlisle W. Baskin* (Central Places in Southern Germany, *Englewood Cliffs, N.J.: Prentice-Hall), Ullman's seminal contribution has been largely neglected.*

The addendum to this article, entitled "Comparison of the Christaller Central Place and the Rank-Size Grouping of Cities," was written initially by Ullman in 1958. A later version, included here, was coauthored by Edward Ullman and Ronald R. Boyce in 1967.

6. A section of the International Geographical Congress at Amsterdam in 1938 dealt with "Functional Relations between City and Country." The papers are published in volume 2 of the *Comptes rendus* (Leiden: E. J. Brill, 1938). A recent American study is C. D. Harris, "Salt Lake City: A Regional Capital" (Ph.D. dissertation, University of Chicago, 1940). Pertinent also is R. E. Dickinson, "The Metropolitan Regions of the United States," *Geographical Review* 24 (1934):278–91.

7. *Die zentralen Orte in Süddeutschland* (Jena, 1935); also a paper (no title) in *Comptes rendus du Congrès internationale de géographie Amsterdam* 2 (1938):123–37.

8. This does not deny the importance of "gateway" centers such as Omaha and Kansas City, cities located between contrasting areas in order to secure exchange benefits. The logical growth of cities at such locations does not destroy the theory to be presented (cf. R. D. McKenzie's excellent discussion in *The Metropolitan Community* [New York, 1933], pp. 4 ff.).

9. Cf. Petrie's statement about ancient Egypt and Mesopotamia: "It

has been noticed before how re-markably similar the distances are between the early nome capitals of the Delta (twenty-one miles on an average) and the early cities of Mesopotamia (averaging twenty miles apart). Some physical cause seems to limit the primitive rule in this way. Is it not the limit of central storage of grain, which is the essential form of early capital? Supplies could be centralized up to ten miles away; beyond that the cost of transport made it better worth while to have a nearer centre" (W. M. Flinders Petrie, *Social Life in Ancient Egypt* [London, 1923; reissued, 1932], pp. 3–4).

10. C. J. Galpin, *Social Anatomy of an Agricultural Community* (University of Wisconsin Agricultural Experiment Station Research Bulletin 34, 1915), and the restudy by J. H. Kolb and R. A. Polson, *Trends in Town-Country Relations* (University of Wisconsin Agricultural Experiment Station Research Bulletin 117, 1933); B. L. Melvin, *Village Service Agencies of New York State, 1925* (Cornell University Agricultural Experiment Station Bulletin 493, 1929), and *Rural Population of New York, 1855–1925* (Cornell University Agricultural Experiment Station Memoir 116, 1928); Dwight Sanderson, *The Rural Community* (New York, 1932), esp. pp. 488–514, which contains references to many studies by Sanderson and his associates; Carle C. Zimmerman, *Farm Trade Centers in Minnesota, 1905–29* (University of Minnesota Agricultural Experiment Station Bulletin 269, 1930); T. Lynn Smith, *Farm Trade Centers in Louisiana 1905 to 1931* (Louisiana State University Bulletin 234, 1933); Paul H. Landis, *South Dakota Town-Country Trade Relations, 1901–1931* (South Dakota Agricultural Experiment Station Bulletin 274, 1932), and *The Growth and Decline of South Dakota Trade Centers, 1901–1933* (Bulletin 279, 1938), and *Washington Farm Trade Centers, 1900–1935* (State College of Washington Agricultural Experiment Station Bulletin 360, 1938). Other studies are listed in subsequent footnotes.

11. See August Lösch, "The Nature of the Economic Regions,"

for a surrounding area are termed "central" functions by Christaller, and the settlements performing them "central" places. An industry using raw materials imported from outside the local region and shipping its products out of the local area would not constitute a central service.

Ideally, each central place would have a circular tributary area, as in von Thünen's proposition, and the city would be in the center. However, if three or more tangent circles are inscribed in an area, unserved spaces will exist; the best theoretical shapes are hexagons, the closest geometrical figures to circles which will completely fill an area (fig. 11.1).[11]

Christaller has recognized typical-size settlements, computed their average population, their distance apart, and the size and population of their tributary areas in accordance with

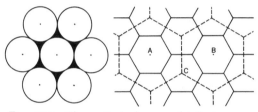

Fig. 11.1. Theoretical shapes of tributary areas. Circles leave unserved spaces, hexagons do not. Small hexagons are service areas for smaller places, large hexagons (dotted lines) represent service areas for next higher-rank central places.

TABLE 11.1

Central Place	Towns		Tributary Areas	
	Distance Apart (km.)	Population	Size (sq. km.)	Population
Market hamlet (*Marktort*)	7	800	45	2,700
Township center (*Amtsort*)	12	1,500	135	8,100
County seat (*Kreisstadt*)	21	3,500	400	24,000
District city (*Bezirksstadt*)	36	9,000	1,200	75,000
Small state capital (*Gaustadt*)	62	27,000	3,600	225,000
Provincial head city (*Provinzhauptstadt*)	108	90,000	10,800	675,000
Regional capital city (*Landeshauptstadt*)	186	300,000	32,400	2,025,000

his hexagonal theory as table 11.1 shows. He also states that the number of central places follows a norm from largest to smallest in the following order: 1:2:6:18:54, etc.[12]

All these figures are computed on the basis of South Germany, but Christaller claims them to be typical for most of Germany and western Europe. The settlements are classified on the basis of spacing each larger unit in a hexagon of next-order size, so that the distance between similar centers in the table above increases by the $\sqrt{3}$ over the preceding smaller category (in fig. 11.1, e.g., the distance from A to B is $\sqrt{3}$ times the distance from A to C). The initial distance figure of seven kilometers between the smallest centers is chosen because four to five kilometers, approximately the distance one can walk in one hour, appears to be a normal service-area limit for the smallest centers. Thus, in a hexagonal scheme, these centers are about seven kilometers apart. Christaller's maps indicate that such centers are spaced close to this norm in South Germany. In the larger categories the norms for distance apart and size of centers appear to be true averages; but variations from the norm are the rule, although wide discrepancies are not common in the eastern portion of South Germany, which is less highly industrialized than the Rhine-Ruhr areas in the west. The number of central places of each rank varies rather widely from the normal order of expectancy.

The theoretical ideal appears to be most nearly approached in poor, thinly settled farm districts—areas which are most nearly self-contained. In some other sections of Germany industrial concentration seems to be a more important explanation, although elements of the central-place type of distribution are present. Christaller points out that Cologne is really the commercial center for the Ruhr industrial district even though it is outside the Ruhr area. Even in mountain areas centrality is a more important factor than topography in fixing the distribution of settlements. Christaller states that one cannot claim that a certain city is where it is because of a certain river—that would be tantamount to saying that if there were no rivers there would be no cities.

III

Population alone is not a true measure of the central importance of a city; a large mining, industrial, or other specialized-function town might have a small tributary area and exercise a few central functions. In addition to population, therefore, Christaller uses an index based on number of

"It was a shock to hear that Louis Wirth had died. I received a lot of stimulus and encouragement from him in my early days and always appreciated it. In fact my article, "A Theory of Location of Cities," was written at his request." [Letter to Norton Ginsburg, 15 May 1952]

"This article introduced Christaller to American thinking. One reason I did this is that I had developed in my own mind a similar system which I was about to write up when Christaller's work appeared. I thus had to content myself with introducing his concepts to the United States, but I did add some comparative data." [Letter to Hans Carol, 18 May 1954]

Southern Economic Journal 5 (1938):73. Galpin, *Social Anatomy of an Agricultural Community*, thought in terms of six tributary-area circles around each center. See also Kolb and Polson, *Trends in Town-Country Relations*, pp. 30–41.

12. Barnes and Robinson present some interesting maps showing the average distance apart of farmhouses in the driftless area of the Middle West and in southern Ontario. Farmhouses might well be regarded as the smallest settlement units in a central-place scheme, although they might not be in the same numbered sequence (James A. Barnes and Arthur H. Robinson, "A New Method for the Representation of Dispersed Rural Population," *Geographical Review* 30 (1940):134–37).

13. In Iowa, e.g., almost all towns of more than 450 inhabitants have banks, half of the towns of 250–300, and 20 percent of the towns of 100–150 (according to calculations made by the author from population estimates in *Rand McNally's Commercial Atlas* for 1937).

14. See particularly the thorough study by B. L. Melvin, *Village Service Agencies*; C. R. Hoffer, *A Study of Town-Country Relationships* (Michigan Agricultural Experiment Station Special Bulletin 181, 1928), data on number of retail stores and professions per town; H. B. Price and C. R. Hoffer, *Services of Rural Trade Centers in Distribution of Farm Supplies* (Minnesota Agricultural Experiment Station Bull. 249, 1938); William J. Reilly, *Methods for the Study of Retail Relationships* (Bureau of Business Research Monographs, no. 4, University of Texas Bull. 2944, 1929), p. 26; J. H. Kolb, *Service Institution of Town and Country* (Wisconsin Agricultural Experiment Station Research Bull. 66, 1925), town size in relation to support of institutions; Smith, *Farm Trade Centers in Louisiana*, pp. 32–40; Landis, *South Dakota Town-Country Trade Relations*, p. 20, population per business enterprise, and pp. 24–25, functions per town size; Zimmerman, *Farm Trade Centers in Minnesota*, pp. 16 and 51 ff.

For a criticism of population estimates of unincorporated hamlets used in many of these studies see Glenn T. Trewartha, "The Unincorporated Hamlet: An Analysis of Data Sources," paper presented 28 December 1940 at Association of American Geographers, Baton Rouge; forthcoming, probably in March number of *Rural Sociology*, vol. 6, 1941.

15. Dickinson, "The Metropolitan Regions of the United States," pp. 280–81.

16. Otto Schlier, "Die zentralen Orte des Deutschen Reichs," *Zeitschrift der Gesellschaft für Erdkunde zu Berlin* (1937), pp. 161–70. See also map constructed from Schlier's figures in R. E. Dickinson's valuable article, "The Economic Regions of Germany," *Geographical Review* 28 (1938):619. For use of census figures in the United States, see Harris, "Salt Lake City," pp. 3–12.

telephones in proportion to the average number per thousand inhabitants in South Germany, weighted further by the telephone density of the local subregion. A rich area such as the Palatinate supports more telephones in proportion to population than a poor area in the Bavarian Alps; therefore, the same number of telephones in a Palatinate town would not give it the same central significance as in the Alps. He claims that telephones, since they are used for business, are a reliable index of centrality. Such a thesis would not be valid for most of the United States, where telephones are as common in homes as in commercial and professional quarters.

Some better measures of centrality could be devised, even if only the number of out-of-town telephone calls per town. Better still would be some measure of actual central services performed. It would be tedious and difficult to compute the amount, or percentage, of business in each town drawn from outside the city, but some short cuts might be devised. If one knew the average number of customers required to support certain specialized functions in various regions, then the excess of these functions over the normal required for the urban population would be an index of centrality.[13] In several states rural sociologists and others have computed the average number of certain functions for towns of a given size. With one or two exceptions only small towns have been analyzed. Retail trade has received most attention, but professional and other services have also been examined. These studies do not tell us actually what population supports each service, since the services are supported both by town and by surrounding rural population, but they do provide norms of function expectancy which would be just as useful.[14]

A suggestive indicator of centrality is provided by the maps which Dickinson has made for per capita wholesale sales of cities in the United States.[15] On this basis centers are distributed rather evenly in accordance with regional population density. Schlier has computed the centrality of cities in Germany on the basis of census returns for "central" occupations.[16] Refinement of some of our census returns is desirable before this can be done entirely satisfactorily in the United States, but the method is probably the most promising in prospect.

Another measure of centrality would be the number of automobiles entering a town, making sure that suburban movements were not included. Figures could be secured if the state-wide highway planning surveys in forty-six states were extended to gather such statistics.

IV

The central-place scheme may be distorted by local factors, primarily industrial concentration or main transport routes. Christaller notes that transportation is not an areally operating principle, as the supplying of central goods implies, but is a linearly working factor. In many cases central places are strung at short intervals along an important transport route, and their tributary areas do not approximate the ideal circular or hexagonal shape but are elongated at right angles to the main transport line.[17] In some areas the reverse of this normal expectancy is true. In most of Illinois, maps depicting tributary areas show them to be elongated parallel to the main transport routes, not at right angles to them.[18] The combination of nearly uniform land and competitive railways peculiar to the state results in main railways running nearly parallel and close to one another between major centers.

In highly industrialized areas the central place scheme is generally so distorted by industrial concentration in response to resources and transportation that it may be said to have little significance as an explanation for urban location and distribution, although some features of a central place scheme may be present, as in the case of Cologne and the Ruhr.

In addition to distortion, the type of scheme prevailing in various regions is susceptible to many influences. Productivity of the soil,[19] type of agriculture and intensity of cultivation, topography, governmental organization, are all obvious modifiers. In the United States, for example, what is the effect on distribution of settlements caused by the sectional layout of the land and the regular size of counties in many states? In parts of Latin America many centers are known as "Sunday towns"; their chief functions appear to be purely social, to act as religious and recreational centers for holidays—hence the name "sunday town."[20] Here social rather than economic services are the primary support of towns, and we should accordingly expect a system of central places with fewer and smaller centers, because fewer functions are performed and people can travel farther more readily than commodities. These underlying differences do not destroy the value of the theory; rather they provide variations of interest to study for themselves and for purposes of comparison with other regions.

The system of central places is not static or fixed; rather it is subject to change and development with changing conditions.[21] Improvements in transportation have had noticeable

17. For an illustration of this type of tributary area in the ridge and valley section of east Tennessee, see H. V. Miller, "Effects of Reservoir Construction on Local Economic Units," *Economic Geography* 15 (1939):242–49.

18. See, e.g., *Marketing Atlas of the United States* (New York: International Magazine Co., Inc.) or *A Study of Natural Areas of Trade in the United States* (Washington, D.C.: U.S. National Recovery Administration, 1935).

19. Cf. the emphasis of Sombart, Adam Smith, and other economists on the necessity of surplus produce of land in order to support cities. Fertile land ordinarily produces more surplus and consequently more urban population, although "the town . . . may not always derive its whole subsistence from the country in its neighborhood . . ." (Adam Smith, *The Wealth of Nations* [New York: Modern Library, 1937] p. 357; Werner Sombart, *Der moderne Kapitalismus*. 2d rev. ed., vol. 1, pp. 130–31 [Munich and Leipzig, 1916]).

20. For an account of such settlements in Brazil see Pierre Deffontaines, "Rapports fonctionnels entre les agglomérations urbaines et rurales: un example en pays de colonisation, le Brésil," *Compres rendus du Congrès internationale de géographie Amsterdam* 2 (1938):139–44.

21. The effects of booms, droughts, and other factors on trade-center distribution by decades are brought out in Landis' studies for South Dakota and Washington. Zimmerman and Smith also show the changing character of trade-center distribution (see note 10 of this paper for references). Melvin calls attention to a "village population shift lag"; in periods of depressed agriculture, villages in New York declined in population approximately a decade after the surrounding rural population had decreased (B. L. Melvin, *Rural Population of New York*, p. 120).

22. Most studies indicate that only the very smallest hamlets (under 250 population) and crossroads stores have declined in size or number. The larger small places have held their own (see Landis for Washington, *Washington Farm Trade Centers*, p. 37, and his *South Dakota Town-Country Trade Relations*, pp. 34–36). Zimmerman in 1930 (*Farm Trade Centers in Minnesota*, p. 41) notes that crossroads stores are disappearing and are being replaced by small villages. He states further: "It is evident that claims of substantial correlation between the appearance and growth of the larger trading center and the disappearance of the primary center are more or less unfounded. Although there are minor relationships, the main change has been a division of labor between the two types of centers rather than a complete obliteration of the smaller in favor of the larger" (p. 32).

For further evidences of effect of automobile on small centers see R. V. Mitchell, *Trends in Rural Retailing in Illinois, 1926 to 1938* (University of Illinois Bureau of Business Research Bulletin, Series 59, 1939), pp. 31 ff., and Sanderson, *The Rural Community*, p. 564, as well as other studies cited above.

23. Smith (*Farm Trade Centers in Louisiana*, p. 54) states: "There has been a tendency for centers of various sizes to distribute themselves more uniformly with regard to the area, population, and resources of the state. Or the changes seem to be in the direction of a more efficient pattern of rural organization. This redistribution of centers in conjunction with improved methods of communication and transportation has placed each family in frequent contact with several trade centers. . . ."

In contrast, Melvin (*Rural Population of New York*, p. 90), writing about New York State before the automobile had had much effect, states: "In 1870 the villages . . . were rather evenly scattered over the entire state where they had been located earlier in response to particular local needs. By 1920, however, the villages had become distributed more along routes of travel and transportation and in the vicinity of cities."

effects. The provision of good automobile roads alters buying and marketing practices, appears to make the smallest centers smaller and the larger centers larger, and generally alters trade areas.[22] Since good roads are spread more uniformly over the land than railways, their provision seems to make the distribution of centers correspond more closely to the normal scheme.[23]

Christaller may be guilty of claiming too great an application of his scheme. His criteria for determining typical-size settlements and their normal number apparently do not fit actual frequency counts of settlements in many almost uniform regions as well as some less rigidly deductive norms.[24]

Bobeck in a later article claims that Cristaller's proof is unsatisfactory.[25] He states that two-thirds of the population of Germany and England live in cities and that only one-third of these cities in Germany are real central places. The bulk are primarily industrial towns or villages inhabited solely by farmers. He also declares that exceptions in the rest of the world are common, such as the purely rural districts of the Tonkin Delta of IndoChina, cities based on energetic entrepreneurial activity, as some Italian cities, and world commercial ports such as London, Rotterdam, and Singapore. Many of these objections are valid; one wishes that Christaller had better quantitative data and were less vague in places. Bobeck admits, however, that the central-place theory has value and applies in some areas.

The central place theory probably provides as valid an interpretation of settlement distribution over the land as the concentric zone theory does for land use within cities. Neither theory is to be thought of as a rigid framework fitting all location facts at a given moment. Some, expecting too much, would jettison the concentric zone theory; others, realizing that it is an investigative hypothesis of merit, regard it as a useful tool for comparative analysis.

V

Even in the closely articulated national economy of the United States there are strong forces at work to produce a central place distribution of settlements. It is true that products under our national economy are characteristically shipped from producing areas through local shipping points directly to consuming centers which are often remote. However, the distribution of goods or imports brought into an area is characteristically carried on through brokerage, wholesale and retail channels in central cities.[26] This graduated division of functions supports a central place framework of settlements. Many nonindustrial regions of relatively uniform land surface have cities distributed so evenly over the land that some sort of central place theory appears to be the prime explanation.[27] It should be worth while to study this distribution and compare it with other areas.[28] In New England, on the other hand, where cities are primarily industrial centers based on distant raw materials and extra-regional markets, instead of the land's supporting the city the reverse is more nearly true: the city supports the countryside by providing a market for farm products, and thus infertile rural areas are kept from being even more deserted than they are now.

The forces making for concentration at certain places and the inevitable rise of cities at these favored places have been emphasized by geographers and other scholars. The phenomenal growth of industry and world-trade in the last hundred years and the concomitant growth of cities justify this emphasis but have perhaps unintentionally caused the intimate connection between a city and its surrounding area partially to be overlooked. Explanation in terms of concentration is most important for industrial districts but does not provide a complete areal theory for distribution of settlements. Furthermore, there is evidence that "of late . . . the rapid growth of the larger cities has reflected·their increasing importance as commercial and service centers rather than as industrial centers."[29] Some form of the central place theory should provide the most realistic key to the distribution of settlements where there is no marked concentration—in agricultural areas where explanation has been most difficult in the past. For all areas the system may well furnish a theoretical norm from which deviations may be measured.[30] It might also be an aid in planning the development of new areas. If the theory is kept in mind by workers in academic and planning fields as more studies are made, its validity may be tested and its structure refined in accordance with regional differences.

24. This statement is made on the basis of frequency counts by the author for several midwestern states (cf. also Schlier, *Die zentralen Orte des Deutschen Reichs*, pp. 165–69, for Germany).

25. Hans Bobeck, "Über einige functionelle Stadttypen und ihre Beziehungen zum Lande," *Compres rendus du Congrès internationale de géographie Amsterdam* 2 (1938):88.

26. Harris, "Salt Lake City," p. 87.

27. For a confirmation of this see the column diagram on page 73 of Lösch ("The Nature of the Economic Regions"), which shows the minimum distances between towns in Iowa of three different size classes. The maps of tradecenter distribution in the works of Zimmerman, Smith, and Landis (cited earlier) also show an even spacing of centers.

28. The following table gives the average community area for 140 villages in the United States in 1930. In the table, notice throughout that (1) the larger the village, the larger its tributary area in each region, and (2) the sparser the rural population density, the larger the village tributary area for each size class (contrast mid-Atlantic with Far West, etc.).

Region	Community Area in Square Miles		
	Small Villages[a]	Medium Villages[b]	Large Villages[c]
Mid-Atlantic	43	46	87
South	77	111	146
Middle West	81	113	148
Far West		365	223

[a]250–1,000 pop.
[b]1,000–1,750 pop.
[c]1,750–2,500 pop.

Although 140 is only a sample of the number of villages in the country, the figures are significant because the service areas were carefully and uniformly delimited in the field for all villages (E. deS. Brunner and J. D. Kolb, *Rural Social Trends* [New York, 1933], p. 95; see also E. deS. Brunner, G. S. Hughes, and M. Patten, *American Agricultural Villages* [New York, 1927], chap. 2).

In New York twenty-six square miles was found to be the average area per village in 1920. "Village" refers to any settlement under 2,500 population. Nearness to cities, type of agriculture, and routes of travel are cited as the three most important factors influencing density of villages. Since areas near cities are suburbanized in some cases, as around New York City, the village-density in these districts is correspondingly high. Some urban counties with smaller cities (Rochester, Syracuse, and Niagara Falls) have few suburbs, and consequently the villages are farther apart than in many agricultural counties (B. L. Melvin, *Rural Population of New York*, pp. 88–89; table on p. 89 shows number of square miles per village in each New York county).

In sample areas of New York State the average distance from a village of 250 or under to another of the same size or larger is about 3 miles; for the 250–749 class it is 3–5 miles; for the 750–1,249 class, 5–7 miles (B. L. Melvin, *Village Service Agencies*, p. 102; in the table on p. 103 the distance averages cited above are shown to be very near the modes).

Kolb makes some interesting suggestions as to the distances between centers. He shows that spacing is closer in central Wisconsin than in Kansas, which is more sparsely settled (J. H. Kolb, *Service Relations of Town and Country* [Wisconsin Agricultural Experimental Station Research Bull. 58 (1923)]; see pp. 7–8 for theoretical graphs).

In Iowa, "the dominant factor determining the *size* of convenience-goods areas is distance" (*Second State Iowa Planning Board Report*, Des Moines, April 1935, p. 198). This report contains fertile suggestions on trade areas for Iowa towns. Valuable detailed reports on retail trade areas for some Iowa counties have also been made by the same agency.

29. U.S. National Resources Committee, *Our Cities—Their Role in the National Economy: Report of the Urbanism Committee* (Washington, D.C.: Government Printing Office, 1937), p. 37.

30. Some form of the central place concept might well be used to advantage in interpreting the distribution of outlying business districts in cities (cf. Malcolm J. Proudfoot, "The Selection of a Business Site," *Journal of Land and Public Utility Economics* 14 (1938), esp. 373 ff).

Addendum: Comparison of the Christaller Central Place and the Rank-Size Grouping of Cities

Both of these systems postulate a regular hierarchy in the number and size of cities with naturally more small ones than large ones. In other words, they assert a regular decrease in size, and increase in number, of places as their "rank" from the largest city increases. Central place theory proposes a step-like downward progression of sizes by *groups* of cities (one largest, followed by two of next size, six next largest, eighteen, etç.). Thus, a series of city-size plateaus of increasing numbers, of smaller size centers (fig. 11.2 and table 11.2) results as rank increases. For example, under a K3 system, each city in the second group will have one-third the population of the largest city, cities in the third group will have one-third the population of the cities in the second largest group, etc.[31] This is based on the "typical" population of the city and its surrounding region. The population within places can be calculated on the basis of this figure.[32]

By contrast, the rank-size rule (also called Auerbach's Law and the Zipf curve) results in a system whereby the population of any given city multiplied by its rank equals the population of the largest city in the country. Stated another way, the population of any ranked city (P_x) is determined on the basis of the population of the largest city in the country as follows: $P_x = P/r_x$ where P is the population of the largest city in the area and r_x is the rank of the city in question, below the largest city. (Turned around: the rank of any city $r_x = P/P_x$.) From this a smooth exponential downward sloping curve of city population results.

In one respect the two systems are consistent. Regardless of whether a central place K3, K4, or K7 network is used, the calculated rank-size population of any given place fits perfectly the *end* ranking city of each central place group (fig. 11.2). However, the rank-size figures are too high in all but these end group cases. For example, given the largest place in an area with a population of 300,000, a central place K3 system yields two second order places of 100,000 each (300,000/3), six third order places of 33,333 each (100,000/3) and eighteen fourth order places of 11,111 each (33,333/3). Note that the progression of total places is 1:3:9:27 (three times each higher order number) but that the number of places in each group is 1:2:6:18.[33]

A more perfect fit would result if the rank-size predictions coincided with the mean population value, rather than the end value, in each central place group. This can be easily

"Before the recent war I had occasion to read your works and review them in this country. Unfortunately, the war prevented my sending you reprints. I am, therefore, somewhat belatedly sending you reprints of three articles which might interest you.

The article on "A Theory of Location for Cities" obviously owes much to you. I wonder if you have changed your ideas any since the time you first developed the central place theory? I believe that I have been quite successful in introducing your theory to this country, inasmuch as there are several people working on similar topics now. I, myself, am interested in both cities and transportation, as you can see from the enclosed reprints."
[Letter to Walter Christaller, 22 April 1949]

31. The K4 and K7 systems, called by Christaller the transportation and administrative systems respectively, have a progression based on four or seven—cities in a K4 system have one fourth the population of the next higher size and a K7, one seventh, etc.

32. Walter Christaller, *The Central Places of South Germany* (1933), translation by C. W. Baskin (1968), p. 67.

33. Christaller, *The Central Places of South Germany*, p. 65.

"Central place is an example of the introduction of a new idea into a field; origins of the idea; conversations with Lösch; easier to have an original idea in English than to translate an old one from the German. Example of simultaneous invention, Lösch, Christaller, Wirth, with Christaller being the first in print. Note Lösch's footnote later—note also Bobeck. Idea unaccepted by geography at first; took five or ten years before it caught on. Idea fitted neither into environmental determinism, *Landschaft*, or areal differentiation school of geography. No mention by Hartshorne in *Nature of Geography*; yet is biggest single idea in geography today probably." [Letter to Edward J. Taaffe, 6 February 1962]

". . . I am struck at the similarity of Christaller's start in Geography and the subsequent development. It is almost identical to my own experience. Thus, my mother gave me a dime store diary which had populations of cities in the front of it— these fascinated me. Similarly, for a birthday I asked for a wall map of the U.S. and later got atlases on which I drew all sorts of imaginary railroad lines. Later I pushed strongly for a functional approach! Still later, this culminated as I told you, in my devising some sort of central place scheme and then finding out that Christaller had done it as well as Lösch." [Letter to Richard Preston, 15 April 1969]

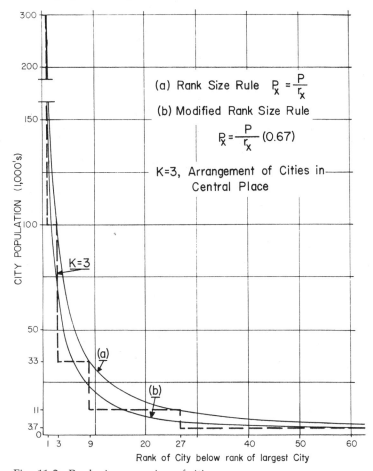

Fig. 11.2. Rank-size grouping of cities

accomplished by multiplying each rank-size population estimate by 0.67—i.e., by reducing each value by one-third (note diagrams).

The reduction by one-third of the rank-size rule thus makes it consistent with the central place scheme in one respect, but in another respect a difference remains: the central place scheme has a step-like arrangement; for example the third category of the K3 central place scheme calls for six places of 33,000, twelve of 18,800 under K4, forty-two of 4,700 under K7, whereas the equivalent rank-size rule (decreased one-third) ranges are six cities from 22,300 to 50,000, twelve cities from 12,500 to 40,200, or forty-two cities from 4,800 to 25,100.

The rank-size rule (especially as reduced one-third) may resemble the actual population figures of places in an area better than the Christaller scheme. This is because the likeli-

TABLE 11.2
Comparison of Central Place City Size Categories with the Rank-Size Rule (Population in 000s)

Rank of Place	Rank-Size Rule Pop	Rank-Size Rule Mod.[a]	Central Place Systems K3 Order* Pop	Central Place Systems K4 Order* Pop	Central Place Systems K7 Order* Pop	Berry-Garrison[b] Central Place K3 Order* Pop
1	300	300	(1) 300	(1) 300	(1) 300	(1) 300
2	150	100.5	100	75	42.9	100
3	100	67.0	(2) 100	(2) 75	42.9	(2) 100
4	75	50.3	33.3	75	42.9	100
5	60	40.2	33.3	18.8	(2) 42.9	33.3
6	50	33.5	33.3	18.8	42.9	33.3
7	42.9	28.7	(3) 33.3	18.8	42.9	33.3
8	37.5	25.1	33.3	18.8	4.7	33.3
9	33.3	22.3	33.3	18.8	4.7	(3) 33.3
10	30.0	20.1	11.1	18.8	4.7	33.3
11	27.3	18.3	11.1	(3) 18.8	4.7	33.3
12	25.0	16.7	11.1	18.8	4.7	33.3
13	23.1	15.5	11.1	18.8	4.7	33.3
14	21.4	14.4	11.1	18.8	4.7	11.1
15	20.0	13.4	11.1	18.8	4.7	11.1
16	18.8	12.5	11.1	18.8	4.7	11.1
17	17.6	11.8	11.1	4.7	4.7	11.1
18	16.7	11.2	11.1	4.7	4.7	(4) 11.1
19	15.8	10.6	(4) 11.1	4.7	(3) 4.7	11.1
20	15.0	10.0	11.1	4.7	4.7	11.1
21	14.3	9.6	11.1	4.7	4.7	11.1
22	13.6	9.2	11.1	4.7	4.7	11.1
23	13.0	8.7	11.1	4.7	4.7	11.1
24	12.5	8.4	11.1	(4) 4.7	4.7	11.1
25	12.0	8.0	11.1	4.7	4.7	11.1
26	11.5	7.7	11.1	4.7	4.7	11.1
27	11.1	7.4	11.1	4.7	4.7	11.1
28	10.7	7.2	3.7	4.7	4.7	11.1
29	10.3	6.9	3.7	4.7	4.7	11.1
30	10.0	6.7	3.7	4.7	4.7	11.1
31	9.7	6.5	3.7	4.7	4.7	11.1
.	.	.	.	.	.	.
40	7.5	5.0	(5) 3.7	4.7	4.7	11.1
41	7.3	4.9	3.7	4.7	4.7	3.7
.	.	.	.	.	.	.
64	4.7	3.1	3.7	4.7	(4) 0.8	(5) 3.7

*Number refers to order of place, e.g., (1) is highest order, (2) is next highest order, etc.

[a]Where population of any place equals the population of the largest place in the area, in this case 300,000, divided by the rank of any given place multiplied by 0.67.

[b]See Brian J. L. Berry and William L. Garrison, "Alternate Explanations of Urban Rank-Size Relationships," *Annals of the Association of American Geographers* (March 1958), p. 86. This method is erroneously applied here, since it assumes 3 second-order cities, rather than two, which produces an increase to 9 rather than 6, etc.. It fits the rank-size rule better than the correct application and thus perhaps accounts in part for their conclusion that "no great difference exists between Zipf and Christaller."

"Now it is quite true that Mark Jefferson was remarkably prescient, and I wouldn't be surprised that he had some sort of "central place" theory in mind, but I doubt whether this statement was any more than accidental. In the first place, reading it in the context in which he was writing the article, leads me to believe that Jefferson may have had a definition of "central" in mind that was more akin to agglomeration as opposed to dispersion. Christaller on the other hand, I believe, had a more rigorously geometrical connotation to "central." Incidentally, I also felt that I had introduced the term "central place" to America since I was the first to translate Christaller's *Die zentralen Orte* to literal English. Now I might have read Mark Jefferson before, but I doubt whether his incidental use of the term registered with me. In any case, I am quite certain that Christaller never read Jefferson since he read little if any English at that time. Although I have not checked his 1933 book, in a recent paper written last year, a few months before he died and published in a German geographical journal, Christaller described how he got the "central place" idea and then indicates that August Lösch later published the same thing, but had developed it independently. Christaller sums up by saying the idea was in the air; I know it was because I also independently had it before reading Christaller. . . ."
[Letter to Geoffrey J. Martin, 24 April 1969]

hood of finding a group of cities of exactly the same size is remote, if for no other reason than the changing character of communities at any one time.

Even more important, however, Christaller's rule applies to a logical process, the central place function alone, and not to specialized functions (mining, manufacturing, resort, etc.), which tend to be more unsystematically deployed spatially. Therefore, the rank-size rule, as some sort of partial probability scheme, may mirror this sort of relatively meaningless average condition, whereas the central place scheme should be more discriminating in singling out a process, based on logical reasoning. According to this argument, there should be more of a tendency toward a central place, step-like arrangement in relatively non-industrial, unspecialized function, and homogeneous areas, as in Iowa, than in Massachusetts or Washington; however, no region is perfect for the measurement; many cases might be needed to verify it.

Are cities really necessary, or even farms, for that matter? Recent trends almost prompt one to raise this question or at the least to ask what kind. This does not necessarily mean that most people will live nowhere and do nothing (but it might help!). Cities have been growing in size, expanding even more in area, and declining in overall density. Analysis of these developments will bring up to date some of "The Nature of Cities," written in 1944,[1] which emphasized, among other facets, that "the support of a city depends on the services it performs not for itself but for a tributary area. Many activities merely serve the population of the city itself." In this presentation the degree to which a city actually is "supported" by performing services for itself will be measured and related to the size and growth of cities. The second and larger part of this study will analyze the expansion of urban areas and bring up to date the increasing importance of the "multiple nuclei" concept of urban structure first suggested in the earlier study.

THE GROWTH OF CITIES

Not only is rural and much small town population declining absolutely, but the very largest cities appear to be increasing more rapidly than any others. Actual figures for relative growth of metropolitan areas in the United States between 1950 and 1960, however, indicate that small and large have all grown about the same in terms of percentage increase. Rates are: over 3,000,000, 23 percent increase; 1,000,000 to 3,000,000, 25 percent; 500,000 to 1,000,000, 36 percent increase (the largest); 100,000 to 500,000, 26 percent. Still other groupings indicate about the same.

However, if the absolute amount of growth is allocated by groups still another interpretation can be made. For example, the five cities over 3,000,000 had an absolute increase of about 5,000,000. The second group also had an increase of about 5,000,000 but this was spread out over sixteen cities. The next smaller groups had about 5,000,000 increase but the increase was distributed over still more cities. Thus an increasing quantity of United States population was concentrated on the average in each of the largest cities. This, then, is presumably the justification for emphasizing metropolitan growth.

This article is Edward Ullman's presidential address to the Regional Science Association in 1962 (Papers and Proceedings of the Regional Science Association, pp. 7–23). The title is taken from the earlier article entitled "The Nature of Cities" (1945), which Ullman coauthored with Chauncy D. Harris; the topic of the address, however, is much broader and more oriented to intraurban affairs than to interurban spatial structure.

"... there is nothing new in geography save perhaps a return to environmentalism if one wants to state it that way—in my case and Isard's an environmental determinism of just one element: space; for Lautensach just one unique unit type: peninsulas, etc. ... the 'environmentalists' were too general and too naive; perhaps you can state why the 'new' environmentalism is (?) more respectable. I can't." [Letter to S. B. Jones, 23 April 1953]

1. Chauncy D. Harris and Edward L. Ullman, "The Nature of Cities," *Annals of the American Academy of Political and Social Sciences*, November 1945, pp. 7–17.

"If I were to guess why he [Lewis Mumford] was against stadiums it would probably be for the same reasons that I seem to be. The spectacles presented in them are hardly elevating to mankind, and secondly represent a sort of non-participatory type of activity which does not seem to have as high priority as a participatory one, such as hiking in the woods, etc. Then again I suppose I might speculate about stadiums that they are used so little that they may be a waste of money. Finally, I suppose another reason I personally am against them is that they are a commercial venture and yet the public is asked to build them and support them. I think that there might be a better use for public money, if for no other reason than to build playfields, better schools, etc. I feel especially indignant about using public money to build a football stadium when football franchises sell for several million dollars. It seems to me the value of that franchise is directly related to the public building of the stadium; therefore, the public should get the money from those franchises! I feel the same way about taxi franchises. It seems ridiculous that a medallion for New York taxi sells for something like $27,000. On the other hand who ever heard of a symphony orchestra franchise selling for anything?" [Undated letter to Bernard H. Booms]

2. Edward L. Ullman and Michael F. Dacey, "The Minimum Requirements Approach to the Urban Economic Base," *Papers and Proceedings of the Regional Science Association* 6 (1960):175–94.

If, in general, each of the largest cities on the average have been growing somewhat more, what is the explanation? No pat answer is possible but the following three factors may be involved. (1) Mere size attracts size—a mass, gravity effect; the larger the center the more innovators, the more persons who have relatives and friends who are attracted as immigrants, etc. (2) The external economies of larger centers provide a greater range of interdependent specialties and facilities. (3) A relative improvement in internal, urban transit has occurred, primarily because of the short haul advantages of the auto and truck; this latter factor has been particularly significant in the expansion of urban area. Leon Moses suggests that the truck allows suburban factories to develop and thus enables the metropolitan area now to compete with outlying regions by providing not only relatively cheap land, but also urban nearness and access to the scale economies just noted.

All three of these forces presumably are given greater scope to influence growth because of the well known shift from primary, to secondary, and particularly to tertiary activities—toward more processing and consequent lesser orientation of production to resource locations.

SCALE ECONOMIES

What is the evidence for the scale economy factor, which has been mentioned so much recently by Vernon, Hoover, and others? In this connection some new findings will be advanced, indicating the degree to which a city is self-contained—takes in its own washing, if you please—which varies according to size and other particulars.

According to studies which Michael Dacey and I, and others, have made using what we call the "minimum requirements" method, there is, on the average, a definite relationship between size of a city and its degree of self-containment [2] (fig. 12.1). Thus, towns of 10,000 have about one-third of their employment serving internal needs and two-thirds external, for an export-internal or basic-service ratio of about 1:5; cities of 500,000 are about evenly divided, one-half internal, one-half external, etc.

This exponential relationship also fits approximately other logical relationships. When extended downward it crosses 0 percent at about four persons, where it should according to logic, since a family unit can sell nothing to itself; when extrapolated upward, a more dubious procedure, it crosses within about 10 percent of the expected for the United States

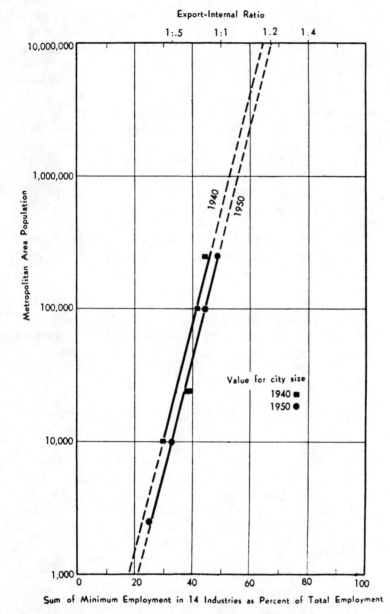

Fig. 12.1. Association of internal (service or non-basic) employment with city size based on minimum requirements method.

". . . my trouble is that I can't get credit for my own balloons, partly because I am so slow to finish off preliminary drafts for publication in journals. I keep seeing my ideas come out in print all the time in someone else's name. This doesn't particularly bother me since it is easier to have ideas than it is to finish them off for publication. . . ." [Letter to Stephen B. Jones, 27 December 1955]

population, as a whole, if one assumes the United States to be about 90 percent self-contained.

If this relationship holds, then the only deduction one can make about the optimum size of cities is that the larger cities are, the more self-contained they are. By extension, on the basis of this measure alone, we tentatively conclude that the larger the city, the more efficient it is, since it can trade more with itself and save transport costs to and from other places. We cannot say there is an optimum size, other than that the larger the better, by this measure.

Table 12.1 indicates some other measures of the gain. Thus the amount of "external" employment "captured" increases, but is the increase really proportionate in effective terms to increase in size of city? For example, is there more "scale economy" gained in increasing from 10,000 to 100,000 than from 1,000,000 to 10,000,000? We do not know and will not know until we know a great deal more about the workings of urban economies.

In any event it does not mean that many metropolises will become multi-million population centers in the next fifty years, first for the obvious reason that the total population of the United States will not be large enough to accommodate many, and secondly because many activities are top hierarchical, one-of-a-kind functions—national headquarters, United Nations, etc. They cannot pyramid in numerous cities. Does the latter consideration mean that one city, New York—or two or three—will become the super giants, as Haig speculated some years ago?[3]

In considering this possibility we encounter other factors—persistence of some resource orientation, whether it be the old ones of minerals and agriculture or the newer role of resort climate, and possible *diseconomies* of scale, or simple lack of scale economies in a significant number of activities, as in government. Foremost among the diseconomies today *may* be environmental limitations—increasing cost of controlling air, and secondly water pollution from large concentrations, although future technology may alter this in unknown ways.[4]

Still other forces are at work in individual cities, as the

3. R. M. Haig, "Toward an Understanding of the Metropolis: Some Speculations Regarding the Economic Basis of Urban Concentration," *Quarterly Journal of Economics* 40 (1926):179–208.

4. A highly urbanized region may also have scale economies. Cf. Chauncy D. Harris, "The Market as a Factor in the Localization of Industry in the U.S.," *Annals of the Association of American Geographers* 44 (1954):315–48; and Edward L. Ullman, "Regional Development and the Geography of Concentration," *Papers and Proceedings of the Regional Science Association* 4 (1958):179–98. See also references to Gottmann and others in footnote 6.

TABLE 12.1

City Size	Approximate % Total Employment		Approximate % of "Remaining" External Employment "Captured" by Increasing City Size Ten-Fold
	Internal	External	
1,000	21	79	—
10,000	32	68	14
100,000	43	57	16
1,000,000	54	46	19
10,000,000	65	35	24

rapid growth of aviation and electronics centered in the attractive climate of Los Angeles, which in turn grows as a second center of the United States in its own, somewhat protected, western territory. With the small number of giant cities over 5,000,000 (three in the United States) it is impossible to single out one common force more important than the individual influences at work on each of the cities. To a lesser degree this is true also of the nineteen cities from 1,000,000 to 5,000,000, which range from Dallas to Seattle to Philadelphia. The individual differences outweigh the similarities, but the scale economy factor would appear to be an underlying force of varying magnitude. Just how this operates and the magnitude of the effect of increasing size, now and in the future, is an explicit question needing further research.

THE INTERNAL EXPANSION OF URBAN AREAS

As our cities grow in size paradoxically their overall densities appear to decline. Suburbs and satellites boom, some fringe areas are by-passed, blight produces a gray area around most of the closer-in parts of central cities, and downtowns decline. This unsettles land values and existing tax bases and alarms powerful groups. The central cities are particularly hit because most of them are unable to expand their city limits. Some conclude that cities are therefore suffering from some unknown disease. There is, however, a logical explanation, already alluded to, related to improved circulation and communication, and particularly to the nature and widespread use of the automobile. Improvements in transportation and communication have benefited short hauls and especially self-loading and unloading commodities like passengers or telephone messages dialed by the individual. These are improvements at metropolitan scale distances.

Before analyzing the forces promoting change, let us attempt to establish what actually has been happening in our cities, a somewhat difficult task both for statistical reasons and because of the recency of the change.

For the country as a whole the census bureau indicates that SMSAs (all central cities and their counties over 50,000) have increased 26 percent in population, but the central cities alone, based on holding city limits constant to 1950, increased only 1.5 percent as compared to 62 percent in the remainder of the metropolitan areas. This is natural; as cities grow, they might be expected to expand in all directions. However, even within the 1950 city limits there is vacant land, especially on the edges, so that actually the innermost portions of cities

"Perhaps my major comment might be summed up by thinking about the future of the German or the European town in comparison with the American: I am wondering whether the differences will become less as the lag between the economic level of the two cultures disappears. I think particularly of the role of the automobile and of increased income. This should show up in farm size, handicrafts, distribution of wholesale activities, . . . dropping of central functions in villages, and disappearance of fairs and general increase in mobility. The fact is that the United States has the automobile, and yet was settled by immigrants from Europe. I don't mean that the cultural differences will disappear, but perhaps it would be instructive to compare the German town with U.S. towns of 30 years ago, or longer. I recall making a study of Streator, Illinois, back about 1939, a city of 15,000, which had a fair amount of wholesaling but which had had even more 20 years earlier because it was a strategic railroad center; it had lost this with the advent of good roads. Much American development in the Middle West responded to the railroad, where I suppose much in Europe antedated the railroad." [Letter to Lutz Hoizner, Jerry E. Mueller, and Edwin J. Dommisse, 23 October 1967]

have declined. In some areas increase in other activities has pushed out residences, but, as will be seen, on an overall basis, probably not even this has compensated for the loss. A net, although unevenly distributed, decline is evident.

For other measures beyond population, it is difficult to obtain data from the census, except for retail trade, and for the office function on a consistent basis it is impossible. To obtain as much consistency as possible a representative group of cities, of all size classes, has been chosen and changes in the central city have been compared to the whole metropolitan area (tables 12.2 and 12.3). The measures were limited to eighteen cities which from 1947 to 1958 had virtually no change in boundaries. Percentage calculations for the average of these cities are given on table 12.2.

These eighteen metropolitan areas and their central cities grew slightly faster than all U.S. metropolitan areas, with the central cities showing a six percent increase compared to one and one-half percent nationally. This probably means, therefore, that the other measures of decline of central city proportion of all metropolitan activity are slightly understated in table 12.1. Also, as noted for population, the measures used probably actually understate the degree of inner decline since the city limits themselves are drawn fairly far out. Finally, the figures are for 1958; later data would show still more relative decline.

Even with these qualifications, relative decline of the central city is apparent. The decrease from 1948 to 1958 in the degree of central city concentration as a percentage of the metropolitan area is shown in the last four figures (28, 29, 30, 31) where Manufacturing in 1958 is 88 percent of 1947, Wholesale Trade 91 percent of 1948, Selected Services 93 percent, and Retail Trade 91 percent, as compared to 84 percent for Population.

The actual percents of population or employment in the central city in 1958 were as follows: population about 54 percent, manufacturing 60 percent, wholesale trade 82 percent, selected services 80 percent, retail trade 72 percent.

It is clear that population leads the way to the suburbs, but jobs are not far behind, especially in manufacturing. Factories appreciate the roominess of the suburb just as much as ranch houses. One-story structures with ample parking are the rule. Walk-up or elevator factories in town are abandoned as soon as conditions permit by most industries, save some with high labor requirements and production processes not sensitive to poor layout or with light weight raw materials and end products. Wholesaling, especially warehousing, should in-

TABLE 12.2

CHANGES IN PROPORTION OF SELECTED ACTIVITIES FOR U.S. CENTRAL CITIES AND
METROPOLITAN AREAS, 1929–60 [a]

Col. No. [b]			Unweighted Mean Percent
1	*Population:* proportion of SMSA located in central city,	1948	64
2		1960	52
3	change in SMSA/Central City concentration,	1960/1948 (1948/1958:84)	81
4	change, SMSA 1960/1948		133
5	change, central city 1960/48		106
6	*Manufacturing:* production workers in central city,	1929	74
7		1939	71
8		1947	70
9		1954	64
10		1958	60
11	establishments in central city,	1947	75
12		1954	69
13		1958	67
14	*Wholesale trade:* paid employees in central city,	1948	89
15		1954	88
16		1958	82
17	establishments in central city,	1948	86
18		1954	83
19		1958	80
20	*Selected services:* paid employees in central city,	1948	86
21		1954	82
22		1958	80
23	establishments in central city,	1954	72
24		1958	69
25	*Retail trade:* paid employees in central city,	1948	79
26		1954	77
27		1958	72
28	*Manufacturing:* change in concentration of prod. workers,	1958/1947	88
29	*Wholesale Trade:* paid employees,	1958/1948	91
30	*Selected services:*	1958/1948	93
31	*Retail trade:*	1958/1948	91

[a] Central cities chosen were those with virtually no boundary change 1947–58: metropolitan area (SMSA) figures adjusted to 1958 area.

[b] Numbers refer to numbers of columns in table 12.3 for individual cities. Sources: U.S. Census of Manufactures 1947, 1954, 1958; Census of Business 1948, 1954, 1958; Census of Population 1960; D. J. Bogue, "A Technique for Making Extensive Population Estimates," *Journal of the American Statistical Association* 45 (June, 1950):149–63.

TABLE 12.3

RATIOS OF ACTIVITIES OF SELECTED CENTRAL CITIES TO SMSAs[a]

	1	2	3	4	5	6	7	8	9	10	11	12	13	14	15	16	17	18	19	20	21	22	23	24	25	26	27	28	29	30	31
Buffalo	56	41	73	124	93	60	50	48	43	42	69	62	61	87	85	80	84	80	75	81	72	70	60	55	70	65	59	88	92	86	84
Chicago	66	57	86	115	99	74	72	70	65	59	83	75	71	88	87	79	88	83	77	85	81	76	71	63	70	70	64	84	90	89	91
Cleveland	65	49	75	123	93	89	86	83	70	68	87	80	76	93	93	87	92	87	81	88	84	83	70	69	78	74	69	82	94	94	88
Detroit	61	44	72	129	94	75	58	60	49	49	69	56	52	86	77	74	87	75	73	86	78	75	66	59	71	65	56	82	86	87	79
Philadelphia	57	46	81	122	98	66	61	71	56	55	71	65	62	89	82	77	85	75	70	77	75	72	62	57	69	63	58	90	87	96	84
St. Louis	52	36	69	126	88	70	71	71	63	56	78	71	68	—	86	80	80	73	70	77	76	73	58	55	70	62	59	79	80	95	84
Akron	68	57	84	125	103	89	82	82	81	67	71	64	59	87	92	85	88	86	82	91	88	82	74	69	81	78	74	82	98	90	91
Miami	53	31	58	209	124	79	77	74	70	50	72	66	57	94	79	72	93	77	70	73	46	45	60	54	66	63	55	68	77	62	83
New Orleans	85	72	85	131	111	78	77	77	68	61	89	83	79	95	91	88	93	90	87	96	90	90	86	82	93	90	86	79	93	94	92
Portland	54	45	83	126	107	68	63	63	60	58	68	56	59	81	91	89	78	85	84	89	84	85	73	68	79	79	77	92	110	96	97
Rochester	69	54	78	123	97	92	94	95	94	92	90	87	84	95	95	92	93	92	91	97	93	87	88	84	85	90	82	97	97	90	96
Syracuse	64	51	80	131	105	76	71	47	43	40	55	49	49	84	79	71	76	74	71	82	78	69	54	51	67	65	60	85	85	84	90
Baton Rouge	79	66	84	145	121	18	17	28	23	37	57	77	83	87	94	92	88	95	93	89	98	95	97	89	91	96	94	132	106	107	103
Des Moines	79	78	99	122	121	91	95	86	77	69	93	84	85	—	96	94	91	93	92	98	96	95	92	90	92	95	92	80	94	97	100
Erie	62	55	89	121	107	67	69	54	56	57	70	67	64	90	90	87	83	82	79	89	85	83	72	67	82	81	77	106	97	93	94
Flint	62	53	85	147	125	96	99	98	80	68	62	56	65	—	89	54	82	83	79	89	90	89	69	71	85	83	79	69	54	100	93
Salt Lake City	67	49	73	150	111	70	61	70	71	72	84	80	77	92	95	93	91	90	90	89	90	91	78	81	85	87	85	103	101	102	100
South Bend	59	56	95	122	115	82	81	80	81	73	74	66	66	83	77	81	81	79	78	77	81	83	73	71	81	79	75	91	98	108	93
Average	64	52	81[b]	133	106	74	71	70	64	60	75	69	67	89	88	82	86	83	80	86	82	80	72	69	79	77	72	88	91	93	91

[a]Numbers at top of columns are identified on table 12.2.

[b]Adjusted to 84 for 1958.

creasingly join manufacturing in low density structures, although its traditional nature, and especially greater market within the city, in contrast to manufacturing, probably explains the greater urban concentration up to now.

The number of individual establishments in manufacturing are more concentrated in the city than is employment, indicating that the larger, more self-contained industries, requiring more space, have led the way to the suburbs. In wholesale trade the reverse appears, indicating probably relatively more warehouse, low labor activity in suburbs.

These general trends in themselves are not conclusive evidence of expansion and lower density, but several studies of individual cities, among them New York, Chicago, Boston[5] indicate the dispersal as well as the classic example of Los Angeles, a city which has grown up in the recent period.

REASONS FOR
THE PRESENT AND FUTURE REARRANGEMENT OF CITIES

Before citing other evidence of the expansion and rearrangement of cities, let us examine the fundamental changes in background which have made this possible. As is well known, improvement in transportation and circulation has changed the nature of urban space, allowing greater distances to be covered[6] and particularly the development of *favored sites*—parts of the city more on the basis of their intrinsic natural and cultural characteristics, and less because of their location or situation. For example, before the automobile, some poor water recreation areas reached by streetcar or train on the edges of cities were very popular. Since the automobile has taken over, these nearby areas, if of poor quality, have declined drastically and visitors travel up to one hundred or two hundred miles to new impoundments or natural water bodies with better water and scenery, or build swimming pools. In this case, both the opportunity to travel and the ability to pay for something better in our increasingly affluent society have created a change.

Thus the stage is set for urban areas. First, as is well known, provision of street cars and mass transit enabled cities to expand, especially out along radial corridors; the volume required for this type of transport tended to focus on one large center—downtown. With the widespread use of the automobile, not dependent on large volume, the interstices could also be served which provided access to enormous additional amounts of land on the expanding circumference. The area of a circle increases by πr^2, which means, for exam-

5. Perhaps the most significant finding along these lines of the New York Regional Study is buried in a footnote added after the study was completed and using last-minute, 1960 census returns: ". . . the tendency to fill up the previously bypassed land of the inlying counties does not appear to be quite as strong as our projection assumes. . . . In general the dispersive population forces in the region seem even stronger than those built into our model" (Raymond Vernon, *Metropolis 1985* [Cambridge, Mass.: Harvard University Press, 1960], footnote, p. 222).

Examples of two earlier but recent quantitative studies proving the shift in urban structures are: John R. Hamburg and Robert Sharkey, "Chicago's Changing Land Use and Population Structures," *Journal American Institute of Planners* 26 (1960):317–23; and *A Report on Downtown Boston*, Greater Boston Economic Study Committee, 1959.

6. Cf. the prophetic statement of H. G. Wells in "The Probable Diffusion of Great Cities" in *Anticipations* (London, 1901) where, in discussing urban growth promoted by improved methods of transport, he says, "It is not too much to say that before [2000] the vast stretch of country from Washington to Albany will be all of it available to the active citizen of New York and Philadelphia. This does not for the moment imply that cities of the density of our existing great cities will spread to these limits"; quoted by K. C. Edwards, "Trends in Urban Expansion," *Advancement of Science* 62 (September 1959):60.

Jean Gottmann (*Megalopolis*, New York: Twentieth Century Fund, 1960) eloquently describes the human geography of the whole area from Boston to Washington as one unit. Norton S. Ginsburg, "The Dispersed Metropolis: The Case of Okayama," *Toshi Mondai* (Municipal Problems, in Japanese), June 1961, pp. 67–76, equally eloquently proposes a new type of city based on several centers and improved transportation.

Some less careful enthusiasts have overplayed the urban explosion however, partly sparked by a change

ple, that doubling the distance from the center increases the area four times.

Most of the inventions in communication also seem to favor a more open pattern. The telephone with its postage stamp rate over wide areas freed dependence on messengers, the movie made it possible to bring entertainment into the communities and neighborhoods from downtown, and the TV now brings it into the individual house and makes the home even more independent of other localities in the city.

What is happening in cities can be compared to what happened to world land use in the nineteenth and twentieth centuries when improved transport enabled distant fertile lands to produce for the world market and in the process compete with less fertile lands nearer the market. Thus, the steamship and railway brought agricultural products to Europe from fertile prairies in America, Argentina, or Australia and either forced abandonment or drastic alteration of agriculture in many less fertile lands in the European market. Thus the present subsidy to European agriculture might be compared to the subsidy to cities through redevelopment programs, although no value judgment is implied.

Cities might thus initially be compared to the von Thünen model of land use around a city, with intensity generally decreasing as distance increased from the central market. Urban transportation, especially the automobile, removes much of the handicap of distance just as the steamship and railway did for the world's regions.[7]

One might thus paraphrase and add to some well known economic principles by coining a new law of *urban expansion* and *specialization* as follows:

As urban transport improves, cities not only can expand in area, but the range or location choice is widened; the more desirable sites wtihin a city can be reached and developed according to their intrinsic advantages.

The second part of the generalization, relating to *site* qualities rather than *situation* qualities, as geographers would define them, is just as important as the first, or expansion part of the law. The monopoly quality of close-in urban locations is weakened.

Even in parts of Europe the same phenomenon is occurring as witness a statement in 1960 by Dr. Aage Aagesen of the Geography Department of the University of Copenhagen:[8]

The intensive urbanization which has developed in proximity to the railway stations seems to have been transformed into a more general, less pronounced urbanization of more extensive areas; this is a natural result of the fact that the importance exercised by

in definition of metropolitan areas in 1950 by the U.S. Bureau of the Census from a minor civil division basis to a county basis. When mapped it appeared as though urbanization had taken a gigantic leap into the countryside. Actually open country still surrounds all major metropolises even on the eastern seaboard of the United States, even though the built-up area and ribbon development, much of it low density, has spread greatly. Cf. Lester E. Klimm, "The Empty Areas of the Northeastern U.S.," *Geographical Review* 44 (1954):25–45. What has happened is more to be measured by invisible indicators in the landscape: commuting, shopping and other trips, telephone calls, TV, etc., spreading out and beyond suburbia and exurbia. (For an example of quantitative indicators of this see: Edward L. Ullman, Ronald R. Boyce and Donald J. Volk, *The Meramec Basin*, [St. Louis, Mo.: Washington University Press, 1962], chap. 1.)

7. Homer Hoyt anticipates me somewhat in this interpretation (as he constantly does) in "Changing Patterns of Land Values," *Land Economics* 36 (1960):114.

8. Aage Aagesen, "The Copenhagen District and its Population" (paper presented to Symposium on Urban Geography, Lund, Sweden, August 1960), published in *Geografisk Tidsskrift* 59 (1960):204–13 (citation on p. 210).

motor-cars and other motor-vehicles on the daily transport is constantly increasing. Another consequence is that there are almost no limits to the choice of residence; this allows preference to be given to *esthetic* considerations by choosing the site in coastal regions, in undulating land, at the edge of a wood or of a lake. A combination of these factors has caused the expansion of the Copenhagen district toward the north, in the sub-glacial streamtrenches of North Zealand filled with lakes and woods. To the west and to the southwest of Copenhagen, in a flat and fertile moraine-land, the relief of the landscape is far from being as attractive and, therefore, has not invited an expansion of the same dimensions.

The same occurs in American cities where waterfront property, as on Lake Michigan in Chicago or Lake Washington in Seattle, is sought, or attractive wooded hill lands in part draw high class residence as in western St. Louis or northwestern Washington, D.C. Likewise, close in hilly sites are by-passed by factories in favor of outlying, level lands.

Thus specialization on the basis of natural site qualities occurs, whereas a hundred years ago, before the street car or auto, close-in Back Bay in Boston was filled in for high class residential use, or centrally located Nob Hill in San Francisco was built up in mansions and Leland Stanford reportedly got cable car service, an invention of the time uniquely fitted to serve hills. Today many, if not most, of these residents have moved to more spacious sites in the suburbs. Thus different natural factors may apply to urban sites than to rural areas, such as scenically attractive land for high grade housing or level land for factories rather than fertile land for crops.

Urban sites, however, for various reasons, probably cannot be rated so much on their natural characteristics as rural lands, but rather more because of certain man-made or *cultural* attributes. The result is a *push-pull* relationship.

First, close-in locations generally are relatively unattractive because of smoke, noise, traffic, crime, and other well-known attributes of crowding.

Secondly, closer in lands may be by-passed by new building for two principal reasons: (1) the generally smaller size of parcels close-in compared to large outlying tracts suitable for large subdivisions and the lower cost mass building techniques of today, and (2) the greater cost of acquiring old structures and paying high land prices near the center as opposed to using raw land farther out.

The cost of acquiring close-in sites may run from $100,000 to $200,000 and more per acre[9] as compared to $10,000 to $25,000 per acre for outlying land. As a result, few one-story or even two-story structures can afford costs of close-in sites, whether for house or factory; at the same time the demand for

9. Raymond Vernon, "The Economics of the Large Metropolis" in "The Future Metropolis," *Daedalus* 90 (Winter 1961):44.

multi-story apartments or other intensive uses is simply not great enough to cover all the gray areas. As a further result the government must subsidize redevelopment, contributing two-thirds or more and the local government the remainder to get site costs down to competitive levels. Even so, the temptation is to build to high densities, which in the past has produced high-rise, low-income housing, in many cases of dubious attractiveness.

Further, anywhere in the city it appears that low density—that is, two-story group houses—are the cheapest way to house people because of lower construction costs, lack of elevators, etc. Even in England this is claimed.[10] As Hans Blumenfeld notes, the cheapest cost building in a country is apt to be the type which is built most.[11] The higher standard buildings may last somewhat longer, but even the average annual payments do not appear to be significantly less.

It is argued that cost of utilities—sewers, water, electricity—is higher if dispersed building is allowed on the fringes. As Lovelace remarks, "The underground system of sewers and water mains is about all that is holding (the city) together,"[12] Even this is questionable, as Lovelace also notes.

Cheaper methods of lagoon sewage treatment or small package plants have been developed for small subdivisions, septic tanks at low densities are suitable on many soils, and even farmhouses have electricity and telephone at not excessive rates. It is true that new schools and other community facilities may have to be built, but these may replace similar facilities close-in which have outlived their usefulness. One-story schools, requiring more land, are preferred to the old urban two- or three-story structures with inadequate playgrounds.

As a concrete example, Lovelace points out that much of southwestern Michigan outside the cities is developed for low density, non-farm uses in an area of sandy soils with high water table so that sewers and water mains are not required.[13] This illustrates graphically a natural site advantage which can now play a role with cheap transportation. Areas unsuitable for septic tanks can be skipped over.

Furthermore, low-density sprawl on the fringes of a city is not unattractive simply because it is low density, but rather because of the way it is done with ribbon development, removal of trees, growth of junk yards, and the like. It is not the low density itself that is to blame. Restraining cities to dense, contiguous settlement is not the only answer, nor even the best answer to unsightly sprawl. Sprawl does, however, produce some obvious inefficiencies.

10. Cf. Myles Wright, "Further Progress" in *Land Use in an Urban Environment* (Liverpool, 1961), p. 251: "The two story house on new land is still the cheapest form of development in Britain."

11. Hans Blumenfeld, *Urban Land*, vol. 21, no. 7, August 1962.

12. Eldridge Lovelace, "Urban Sprawl Need Not Be a Tragedy," *Landscape Architecture* 51 (1961):230–31.

13. Lovelace, "Urban Sprawl."

THE CENTRAL BUSINESS DISTRICT

The core of the city is generally declining relatively and in many cases absolutely. The best data indicating these trends are for retail sales. The top part of table 12.4 shows change in CBD sales and SMSA sales in terms of constant value dollars from 1948 to 1958. The decreases for CBDs range from 16 percent for cities over 1,000,000 down to about 10 percent for those from 100,000 to 250,000. At the same time the remainder of the SMSAs outside the CBDs were increasing from 33 to 64 percent. The lower portion of table 12.3 shows what percentage CBD sales are of total SMSAs. Note the decline from 16 percent to 10 percent for those over 3,000,000, from 26 to 15 percent for 1,000,000 to 3,000,000, etc. Note also the lower percentage of total SMSA sales in the CBD in the larger cities, as would be expected, ranging from 9.6 percent in the largest group to 32 percent in the smallest. Pre-war, the only firm figure we had was the special census under Proudfoot for Philadelphia which reported 37.5 percent in 1937.

These figures show the effect of the construction of large branch department stores and shopping centers and the general movement of shopping to customers. If much of the retail trade leaves downtowns what will replace it?

Before attempting to answer this question, two fundamental points about downtowns should be noted:

1. Most large cities have developed on water and have grown more in one direction than another so that the central business district is not now centrally located in many cities. As a result it loses sales and economic activities as cities grow away from it.[14] Street grids and mass transport focusing on the central business district mitigated this handicap in the past and the construction of radial superhighways to downtown will probably help overcome it to some extent in the future, especially if the parking problem can be solved.

2. Even more serious than the off-center location, in many cases, is the surrounding of the CBD by the low income, blighted, "gray area" of cities. Redevelopment, therefore, in many cases is pushed in part as a means of providing customers. In addition, a market for high and medium income apartments can be developed around downtowns, especially as older people with grown children come onto the market, as well as a new wave of post-high school and young college graduates. This market in most cities, however, does not appear large enough to affect a significant change. Probably a larger natural apartment market for retired persons exists in suburbs and other centers.

14. Cf. Ronald R. Boyce's forthcoming study suggesting this point of the relation of CBD retail sales to CBD centrality. Also note William Weismantel, "A Multicenter Transportation Plan," *Washington University Law Quarterly*, June 1962, pp. 310–37, for an excellent discussion of St. Louis' growth patterns in relation to transportation.

TABLE 12.4

CENTRAL BUSINESS DISTRICT SALES DATA

I. *Changes in Retail Sales CBD and Metropolitan Areas, 1948–58 Adjusted to 1948 Dollars and for 1960 SMSAs.*

| | All Retail Sales | | | | | | Women's Specialty Stores (Clothing) | | | | | |
| | Changes in CBD Sales | | | Changes in SMSA Sales (Less CBD Sales) | | | Changes in CBD Sales | | | Changes in SMSA Sales (Less CBD Sales) | | |
SMSA Population (1960)	1948–54	1954–58	1948–58	1948–54	1954–58	1948–58	1948–54	1954–58	1948–58	1948–54	1954–58	1948–58
3,000,000 or more (5 cities)	−11.6%	−5%	−16%	+21%	+11%	+33%	− 2.1%	+11.5%	+ 9.3%	+13%	+ 4%	+ 17%
1,000,000 to 3,000,000 (14 cities)	− 7.8%	−7.8%	−16.3%	+31%	+17%	+50%	−16.8%	− 5%	−21%	+12%	+32%	+ 48%
500,000 to 1,000,000 (25 cities)	− 7%	−8%	−14.4%	+38%	+23%	+64%	−11.7%	− 6.7%	−17.5%	+16%	+29%	+ 49%
250,000 to 500,000 (32 cities)	− 4.3%	−6.7%	−10.6%	+35%	+18%	+55%	−10.3%	−13.7%	−23%	+60%	− 3%	+ 55%
100,000 to 250,000 (14 cities)	− 2.3%	−7.1%	− 9.7%	+30%	+18%	+50%	−10%	−10%	−20%	+101%	+13%	−118%
U.S. Average	− 6.6%	−7%	−13.4%	+30.4%	+16.8%	+51%	−10%	− 4.8%	−14.4%	+40.2%	+15%	+ 57%

SOURCE: Calculated from U.S. Census of Business, 1958, 1954, *Central Business District Statistics.*

II. *CBD Retail Sales as Percentage of SMSA Sales, 1948–58.*

| | All Retail Sales, % of SMSA Sales in CBD | | | Women's Specialty Stores (Clothing), % of SMSA Sales in CBD | | |
SMSA Population (1960)	1948	1954	1958	1948	1954	1958
3,000,000 or more (5 cities)	15.6%	11.4%	9.6%	33.8%	27.4%	26.8%
1,000,000 to 3,000,000 (14 ciites)	26%	18.8%	15.4%	58.8%	47.5%	41.1%
500,000 to 1,000,000 (25 ciites)	34.3%	24.3%	19.7%	78.3%	65%	55.3%
250,000 to 500,000 (32 cities)	38.7%	28.5%	24.4%	84.8%	75.6%	70.8%
100,000 to 250,000 (14 cities)	44.5%	37.2%	32.1%	91.7%	81.8%	77.8%
U.S. Average	31.8%	24%	20%	69.5%	59.5%	54.4%

SOURCE: Calculated from U.S. Census of Business, 1958, 1954, *Central Business District Statistics.*

NOTE: From 1958–63, recently released Census figures indicate that CBD sales have continued to decline and at a slightly greater rate. (See Edward L. Ullman, et al., *Recent Changes in Central Business District Sales,* Washington Center for Metropolitan Studies, 1965.)

The remaining large activity for CBDs is the office function. This is growing, and growing particularly in New York which has witnessed a boom in central office and other activities locating there for national control, in part made possible by the airplane. To a degree the same is happening in Washington. For most cities this does not appear so likely. Even Chicago's recent expansion and planned new construction will only result in the same per capita office space as in 1930, although it will help the Loop.[15] Most other cities are worse off.

The unknown question is how much is face-to-face contact—linkages of various kinds—necessary for various functions, especially outside New York City. Many activities apparently do not require it, particularly in insurance and in single-function office buildings.

In some cities, even besides Los Angeles, notably St. Louis, outlying office centers are now starting to develop. Clayton, seven miles west of the CBD and more centrally located in reference to the high income area, has many modern, city-wide or nation-wide office buildings, with rents as much as three times higher than downtown, but with land values only about one-third; Clayton illustrates a location nearer the geographic center of a city as well as closer to executives' homes. Ancillary businesses and social services, including luncheon clubs, have sprung up, although the center is not as large as downtown St. Louis. Executives, however, can still go downtown for luncheon club conferences. They drive to their offices in Clayton, then drive downtown for lunch and return in the afternoon, avoiding all rush hour traffic.

Many activities are downtown just because they are there, or in response to linkages which disappeared years ago. Many could be served better elsewhere. In any case, the average downtown should be greatly improved in order to compete with the greater number of sites now accessible by modern transportation. This will be increasingly difficult in view of the outward movement of housing, retail trade, manufacturing, and other activities which now begin to reinforce each other elsewhere in the city.

It looks as though the central business district may become one of the many centers in a city, in many cases the most important, but a center of much less relative importance than in the past. A logical development would make it the shopping center for the large, low income area around it and an office center on a reduced scale for older activities or smaller concerns needing poor, vacant space or using large amounts of cheap labor. The high grade activities characteristic of the

15. *Urban Land*, April 1961, p. 8.

top hierarchical position of the CBD will abandon it for centers better located to serve the high income areas.[16]

Other centers elsewhere will develop on a regional or specialized basis, strengthening the multiple nuclei generalization suggested in the earlier "Nature of Cities." Conventions and out-of-town visitors will find it increasingly more convenient to locate near the airport which, because of its own space needs locates on the periphery; outlying shopping centers will handle retail trade; large factories and employment centers will be on the outskirts on large tracts of land; special entertainment, educational, cultural, and recreational centers will be scattered all over the city to serve the whole population.

Many have said that a city cannot exist without a heart, the CBD. The metropolis of today and increasingly in the future is not only one city, but a federation of general and special centers. As such it is likely to have several hearts better located than one, and basically will be better off because of reduction in travel time, congestion, and utilization of better sites.

CONCLUSION

The generalizations about urban growth and rearrangement will vary with individual cities because individual natural environments, economic bases, and civic actions vary. Many of the location changes in cities hinge on small margins with inertia and tradition holding many activities in uneconomical, old areas. Identical offices and industries can thrive in central business districts, suburbs, and small towns. They adjust accordingly.

If we were to start over, however, we would not build our cities as they are today. If we were to apply private enterprise depreciation principles to the inner portions of cities we would write them off—just as machinery is scrapped—and throw them away, but where would we throw them?

As a citizen I recognize that the major problem of cities—slums and the gray area—cannot be tolerated. We may well have to eliminate them before we eliminate all the causes, including poverty, ignorance, and racial discrimination against the new arrivals, or the other manifold ills of our society both old and new.

Some might say that the new pattern of our cities is the result of a plot hatched by Detroit, the subdividers, and land speculators. Inflated land values are a part of the "pernicious" process of urban sprawl.[17] The auto does not pay its fair

16. This will eliminate some of the cross hauling now occurring as executives travel from the residential suburbs into the center and workers travel from the center outward to suburban industrial sites.

17. Cf. the thoughtful article by Mason Gaffney, "Urban Expansion—Will It Ever Stop?" *1958 Yearbook of Agriculture* (Washington, D.C.: U.S. Government Printing Office, 1959), pp. 502–22.

share for use of the city and hidden costs are passed on to the public in urban expansion. This may be true, but three points seem germane. (1) The magnitude of the underpayment is probably not enough to result in anything more than a slowing down in the process, even if corrected. (2) Countervailing forces are already deployed on the other side, sparked in part by the threatened decline in land values in the center. Urban redevelopment is subsidized, and priority is given a radial pattern for the interstate highway system focusing on downtown, reflecting old flow patterns, with generally only one circumferential, when some inner or intermediate belts are also required. (3) Even if the whole process is a plot, it is our foreseeable institutional arrangement and as a geographer I see it producing the future expansion—specialization—federation patterns sketched above.

A key question then will be the interrelations between the centers and parts. How much will they benefit from being adjacent, or would separate cities of 100,000 to 500,000 be as good or better? The latter seems unlikely since there are still some specialized services, such as jet aircraft flights, that are better performed for millions than thousands. The problem remains to design cities to take advantage of scale economies and the other advantages of concentration, and at the same time to provide optimum livability.

CHAPTER 13   Problems of the Industrial Landscape: The City and Environmental Quality, Especially Air Pollution Sources and Costs

This article was published in Der Mensch und die Biosphäre *(Verlag Dokumentation, Pullach bei München, 1974), as part of the Proceedings of a UNESCO conference held in Bonn, Germany, in June 1971. With its appearance in the present volume, the paper is for the first time made readily available to American readers.*

The study contains numerous insights into the geographical nature of pollution, particularly in highly industrialized cities. In fact, Ullman was much interested in all aspects of large cities related to their quality as places to live, for example, noise level, physical safety, preservation of views, and visual aesthetics. He had much to say, for instance, on the overhead power lines in Seattle and the growing number of billboards. He was especially interested in air pollution, however, since it had clear spatial dimensions, both as it varied within the city and as it tended to lessen toward the edges of the city. His major concern was always with the effect of pollution on people; therefore, he talked about the amount of "pollution per nose."

1. G. Evelyn Hutchinson, "The Biosphere," *Scientific American* (special issue on the biosphere by several authors) 223 (September 1970):45.

The principal problem associated with the industrial landscape is its spillover effect on man himself. This spillover generally occurs near the industry and since industry is generally in cities, and since cities now house most of the western world's population, pollution is predominantly an urban affair. Air pollution is probably the main concern but water pollution, noise pollution, visual pollution, and solid waste disposal all constitute problems.

Secondary and potential longer range effects might turn out to be critical at some future date. These would be effects on the much larger biosphere of the earth as a whole, as in heating or cooling the atmosphere, but their effect now is minimal and their future consequences still largely speculative, although more research needs to be done.

Not only do urban areas have the most pollution but, in general, the larger the city, the greater is the problem, especially for air pollution.

The biosphere, as conceived by Suess, Vernadsky, and presumably originally Lamarck, is defined as that part of the earth in which life exists. This limits it, according to G. Evelyn Hutchinson, to a short distance above the earth and below the surface of the seas and even eliminates parts of the earth's surface which are too dry, too hot, or too cold to support metabolizing organisms.[1] Three additional special characteristics of the biosphere appear to be: substantial quantities of liquid water, ample supply of energy, and interfaces between liquid, solid, and gaseous states of matter.

Cities thus might be said to represent the most extreme state of the biosphere. For a conference on *man* and the biosphere cities unquestionably also represent the most concentrated segment of man himself. The effects of the industrial landscape thus will be measured directly, but in some cases, indirectly, on man himself, not on *land*, *atmosphere*, or *water*. For air pollution, for example, it is the effect per capita, or per nose. The same amount of pollution on a dense population is more serious than on a sparse population. The total pollution produced thus is generally *less* important than the *concentra-*

tion. If this emphasis on man represents a humanistic approach, the facts cited later indicate it is also a scientific one.

There are of course severe problems of the industrial landscape outside cities, as in the increasing practice of strip mining. Indeed the greatly reduced cost of moving earth—the development of earth moving machinery—has sharply etched urban-industrial characteristics on the landscape in the form of deeper cuts and larger embankments.

Urban-industrial and mining landscapes are generally considered the least attractive and desirable locales in the world, in spite of the fact that they are the richest. One therefore is depressed by the sign at the entrance to Butte, Montana: "Richest Hill on Earth" (and ugliest!).

The following quotation from Professor Jacob Spelt of the University of Toronto characterizes one such classic landscape, the Ruhr, typical, except bigger than many others, including the dreary Jersey Flats opposite New York, or the equally repulsive Calumet-Gary region on the southern margin of Chicago.[2]

A most serious handicap in attracting new industries is that the Ruhr has an unfavourable image. In the minds of the public, the district is identified with coal mining, heavy iron and steel, and a landscape to match. This is a more serious problem for its continued growth than the coal crisis. The Ruhr lacks many amenities compared with other industrial locations in Germany or across the border. Coal mining has moulded and shaped the Ruhr landscape, which in many respects is harsh and ugly. It is a scene characterized by mine heaps, air pollution, destruction of vegetation, land subsidence, lowering of the ground-water table, water pollution, dreary workers' settlements, and so forth . . . [although] the Ruhr does not contain any real slums. Its worst [housing] compares favorably . . . with mining areas of England or France."

There is also a reverse effect of the natural biosphere on the industrial landscape. Industry and cities, like metabolizing life itself, also avoid places that are extremes. Exceptions occur as in industries established to extract minerals from or even created by extreme environments—borax from Death Valley or nitrates from Chile, to say nothing of the need for creating artificial biospheres through air conditioning deep in mines under the surface of the earth.

The principal specific effect of the biosphere, however, appears to be on the sites for industry and cities. Thus industry is almost exclusively confined to flat, low land, in contrast to residences, both because of internal requirements and because transportation, particularly by water and rail, is concentrated in valleys. The natural problems thus include need for draining and filling low wet areas and dangers from flood-

"Strips—I hate strips. Sprawl is okay. I see nothing wrong with low density housing and I've always been amused at the pejorative nature of sprawl. Strips, however, it seems to me, deserve a prize as the ugliest features of America, primarily because of their excessive chaotic and uncontrolled signs, but secondly because of overhead wires. The chief offenders are the gas stations. Their signs are too large. In Europe they are smaller and not as ugly. It is true that strips provide some service particularly useful for people with cars, such as nurseries, lumber yards, etc., many of which are not particularly beautiful, but much of the strip is not of this character. Who cares what kind of gas one buys? The companies change names but they still put large signs up advertising their undifferentiated product." [Letter to Grady Clay, 27 July 1971]

2. Jacob Spelt, "The Ruhr and Its Coal Industry in the Middle 60's," *Canadian Geographer* 13 (1969):8.

"I recognize that knowledge of products and services available is needed—truly informational advertising (such as when do the trains leave, etc.), both to serve the consumer's needs and to allow for economies of scale, but this is not what we get, or if we do, it is too much. TV is a bore. The Sunday New York Times weighs too much. One has to weave through countless ads to read most magazines. Even worse, I suspect policies are affected by a sort of Gresham's law of information. Banal or ugly billboards deface landscapes and city streets (an exception is the beautiful, medieval, matching signs on the old street in Salzburg). Banality and boredom are out of control and opportunity to improve our lives is lost." [Letter to Paul Samuelson, 10 March 1976]

3. T. J. Chandler, *The Climate of London* (London: Hutchinson, 1965). Warren R. Bland, "The Distribution of Air Pollution in the Los Angeles Basin," paper presented to Association of Pacific Coast Geographers, Victoria, B.C., 14 June 1971. *Report for Consultation on the Washington, D.C. National Capital Interstate Air Quality Control Region*, U.S. Public Health Service (c. 1970, mimeo) and similar reports for New York region, and more detailed ones for St. Louis and Nashville. David O. Lutrick, *Mobile Sampling with the Integrating Nephelometer*, master's thesis in engineering, University of Washington, Seattle, 1971.

4. Reid A. Bryson and John E. Kutzbach, *Air Pollution* (Association of American Geographers, Resource Paper no. 2, Washington, 1968), p. 3.

5. H. E. Landsburg, "The Environmental Variations of Condensation Nuclei," *Bulletin of the American Meteorological Society* 18 (1937):172 (as reported in Bryson and Kutzbach, *Air Pollution*, p. 3).

ing, both of which pose substantial costs and dangers. Thus industry itself might be said to be particularly susceptible to natural hazards.

The location of industries and cities in valleys and basins ironically also provides the poorest natural ventilation and hence compounds air pollution. The hill towns of southern Europe were set up initially to avoid the twin dangers in the lowlands of malaria and invaders, whether by land or sea. Unfortunately, they are not very accessible for industry, although their air drainage might be excellent.

AIR POLLUTION

Even though our information is incomplete, air pollution is generally considered to be the major pollution problem of cities and industrial landscapes today, partly because water pollution was attacked seriously earlier and safe drinking water provided. It is harder to purify the air we breathe than the water we drink; water is at least channeled and does not have to be consumed constantly. Its confinement in space and time makes it more amenable to certain types of control, although its movement in concentrations over long distances means that its effects can be transmitted longer distances than air pollution.

Air pollution is the most serious problem because of its concentration near the sources of emission and the location of population at these points. We are beginning to get maps of smoke or air pollution over cities that show a typical gradient of decrease of three (or more) to one, from the center of a city or center of industry to the outskirts of the built-up area.[3]

Still other measures indicate even more of a gradient. Bryson and Kutzbach quoting some earlier studies cite Biel's findings of 17,000 Aitken nuclei condensations (a "dust" measure) in the Vienna woods compared to 190,000 in the business section, a ten-fold increase, and a further twenty times more in the industrial section. (These measures are at twenty-inch heights.)[4] Landsburg in 1937 indicates an "average count" of 147,000 in twenty-eight cases in cities of more than 100,000 and 34,300 in fifteen cases in towns of less than 100,000, and 9,500 in twenty-five countryside cases.[5] Bryson and Kutzbach note many cases of dusty air in the countryside, one of the most extensive being in West Pakistan-Northern India where during the dry season the concentration of dust particles in the air probably averages 500–800 grams per cubic meter, more than is found in the most polluted 10 percent of American cities, and covering a far wider area than any city, although the circulation of the air tends to keep it concentrated in the region.[6]

Bryson and Kutzbach on still different measures indicate considerable dispersion of pollution. In the United States, air on the West Coast coming across the Pacific on the prevailing westerly circulation has excellent visibility except when reduced by fog; by the time it reaches "the midwest the visibility has been reduced to 10 or 15 miles," and by the time it leaves ". . . has been halved again, and often reduced even more."[7] Wind circulation and ventilation over many large areas profoundly affect the spread of pollution. "With no wind air pollution would be a local problem. Only the inhabitants of locally polluted areas would breathe polluted air."[8] Hilst, among others, notes the accumulation of pollution on the downwind side of cities and speculates that if cities are more than one hundred miles long, a condition being approached in some larger megalopolitan areas, concentrations dangerous for community health may be exceeded.[9]

Furthermore, there have been indications of spillovers even across international boundaries in northern Europe in the form apparently of acid rain produced by sulfur emissions in neighboring heavily industrialized areas.

In any case smoke and pollution are confined primarily, but not exclusively, to the built-up area of the city and its downwind side, with greatest concentration in the most densely populated or industrialized parts of the city. If one adds daytime population in industrialized areas and in the commercial cores of cities, the concentration of effect near the centers can be even more marked. A densely populated city also tends to have more pollution per capita than a sparsely populated one, unless there are some special considerations such as the inversion trap situation of Los Angeles. Dilution in the atmosphere is great, but if there is enough pollution it spills at least a short distance beyond the emission area to produce the gradients just noted.

Air pollution varies according to the industrial structure of the city as might be expected, but also with the size of a city, since large cities not only produce much pollution simply from power generation, transportation, and heating, but also because most of them have a fairly large concentration of industry. Thus the cities we think of as being the most polluted are also generally the largest in the world—Tokyo, Los Angeles, Chicago, New York, London, metropolitan areas of 7–14 million population. Most cities of less than 50,000 have only a minimum pollution problem unless they are the site of

6. Bryson and Kutzbach, p. 6.
7. Ibid., p. 1.
8. Ibid., p. 16.
9. G. R. Hilst, "What Can We Do to Clean the Air?" *Bulletin of the American Meteorological Society* 48 (1967):710–13.

a smoky industry. A survey of expert opinion in 1961 in central cities in the U.S. indicate that *all* cities (not SMSAs) over 1,000,000 had major pollution problems and only 10 percent of those in the 25–50,000 category.[10]

COMPOSITION AND SOURCES OF AIR POLLUTION

The "big five" of identified major components of air pollution and their estimated total weight in the United States in millions of tons per year (about 1965) are: particulates (12), sulfur oxides (23), nitrogen oxides (8), carbon monoxide (65), hydrocarbons (15), and other (2), for a total of 125 million tons.[11]

The common way of measuring these pollutants by weight does not give a true picture of their effect; it particularly overstresses carbon monoxide, principally produced by automobiles, whose effect is not nearly as great as its weight. Various indexes have been used to weight this weight based on tolerance levels in man. One such is Pindex (pollution index) developed by Lyndon Babcock in a comprehensive doctoral dissertation finished in 1970 at the University of Washington. This index, like others, is only a beginning until more refined tests are made (col. 2, table 13.1). Another earlier version is a method used by the Russians (col. 4), based on room exposures to pollution levels wherein if carbon

10. "A Study of Pollution—Air: Staff Report to Committee on Public Works, U.S. Senate, September 1963," in *Economics of Air Pollution*, ed. Harold Wolozin (New York, 1966), p. 199.

11. Lyndon Babcock, *Some Air Pollution Implications of Energy Transformation and Use* (Ph.D. dissertation, Department of Civil Engineering, University of Washington, 1970), p. 18. This unpublished dissertation gives an up-to-date and extensive discussion, and makes a significant, original contribution to the problem of air pollution. It also has an extensive bibliography, including the source of the basic pollution composition figures cited: National Academy of Sciences, *Waste Management and Control*, pub. no. 1400, Washington, D.C., 1966. (For another index of pollution see also Bland, "Air Pollution in Los Angeles Basin".)

TABLE 13.1
U.S.A. CORRECTED POLLUTION SOURCE DISTRIBUTIONS

| | Emissions, percent of total | | | |
| | 1 | 2 | 3 | 4 |
Pollution Source	Gross Weight	Pindex only	Urbsim with Pindex	Russian Weights Applied to Col. 1
Transportation (mostly auto)	60	19	34	33
Industry	18	38	13	30
Power Plants	13	29	27	25
Space Heating	6	12	22	10
Refuse Incineration	3	2	4	2

SOURCES: Columns 1, 2, and 3, table 19, Lyndon Babcock, "Some Air Pollution Implications of Energy Transformation and Use." (Unpublished Ph.D. dissertation, University of Washington, 1970.) Column 4, Victor Mote, Department of Geography, University of Washington.

monoxide (CO) has a weight of 1, sulfur oxide (SO_2) is 6, particulate matter 7, hydrocarbons (HC) 3, and nitrogen oxides (NO_x) 12.[12] Still another weighting has been contrived by Warren R. Bland, assistant professor of geography, San Fernando Valley State College in which CO is 1, SO_2 is 2, NO_x is 17, and O_3 is 100. Ozone (O_3) is particularly critical in Los Angeles-type photochemical smog and is also introduced into Babcock's index. It occurs less in northerly cool climates.

Note that transportation goes down from 60 percent by gross weight (col. 1) to 19 percent (col. 2) and industry and power plants go up. The Pindex calculation (col. 2), however, probably understates transportation since most of this category is the automobile and automobile exhausts are at ground level, in contrast to industrial and power stacks. Space heating should also be increased for this reason, and particularly for its location close to population; this same correction should also be made for the automobile; in addition present emission and control equipment for industry and power should be included.

A principal feature of the corrected index in column 3 is that it is based on Babcock's simulated city (Urbsim) which in turn is carefully tailored to give an estimated generalized city characteristic, including locating power plants near the center of the city as they are, and having them burn coal, as they do. Of course for a few cities, such as Seattle and Los Angeles where little if any coal is used, power plants would not be a source of as much pollution. Babcock also locates the industry of his hypothetical city in one sector extending from the center in a pie-shaped (45 degrees) wedge extending to the outskirts of the metropolitan area, with emissions concentrated at points in the central axis of the sector. If industry were closer to the center, as in many old cities, then the industrial percentage would rise; likewise if power plants were outlying, power would not cause as much pollution because it was less centrally located.

The result of these adjustments in column 3 gives what Babcock estimates is probably the fairest allocation of air pollution guilt: transportation 34 percent (if only automobile about 30 percent), industry 13 percent, power plants 27 percent, space heating 22 percent. By comparison, if the Russian weights are applied to the United States gross weights in column 1, the figures are remarkably similar except that industry goes up from 13 to 30 and space heating down from 22 to 11 percent.

In any event, by whatever method, industry and power plants, both parts of the industrial landscape, account for

12. Victor Mote, "Geography of Air Pollution in the U.S.S.R.," Ph.D. dissertation, Department of Geography, University of Washington, 1971.

about 40 to 50 percent of air pollution over American cities, but the automobile for about 30 percent. This is quite different from measurements based primarily on raw weight figures; the United States government, for example, used to indicate that about two-thirds of U.S. air pollution was caused by the automobile; more recent releases have slipped to a figure of 50 percent, not an indication of decline, because autos have been increasing. The actual figure over cities, where pollution is a problem, however is closer to 30 percent in the United States. In Europe, where cars are less used the figures are probably even lower, although since urban populations are denser auto traffic moves more slowly and puts out more pollution per distance and per time and is the most severe problem in many central areas of European cities, as in Rome or Madrid, or capital cities elsewhere in the world.

The findings above generally demonstrate that industry, including power plants, are the principal contributors to air pollution over cities, especially in industrialized areas.

COSTS OF AIR POLLUTION

No good overall estimates of the costs of air pollution have yet been made. Economists, for example, in their earlier and very recent attention to the subject have been bemused by stressing that air pollution is a classic example of externalities, i.e., the cost is passed on to someone else and economic theory does not like this. Time precludes an exhaustive analysis of the problem, but recent studies are beginning to show in persuasive fashion that costs are probably high.

From the health standpoint the traditional view has been that smoking is far more definable and important a cause of morbidity and mortality than air pollution. It may still be, but many studies have shown significant correlations to air pollution in non-smokers and between smokers in polluted and unpolluted areas. A recent thorough review of the literature in England and the United States by Lave and Seskin indicates that a 50 percent reduction in pollution levels in major urban areas in the United States would save more than two billion dollars a year, half for respiratory and the remainder for cardiovascular and cancer illnesses and deaths, based on earnings foregone.[13] This method they indicate is a gross underestimate of the actual amount people pay to keep someone alive now. However the problem is similar to discounted future benefits—a benefit sometime in the future is not as great as one right now. In any case, this cost of certain ills is only a fraction of total benefits even for a 50 percent reduction in air

13. Lester B. Lave and E. P. Seskin, "Air Pollution and Human Health," *Science* 169 (21 August 1970):723–33; reprinted in part in *Ekistics* 185 (April 1971):295–303.

pollution. One robust estimate of the annual health cost of air pollution runs between 14 and 29 billion dollars.[14] The Environmental Protection Agency (EPA) in the *Annual Report* of the Council of Environmental Quality, released only in August 1971, estimates that from medical care and work loss alone the annual cost of air pollution is six billion dollars. If costs from "discomfort, frustration and anxiety were included, these estimates would be greatly increased."[15]

Added to health costs should be material costs of cleaning, painting, and deterioration annually of several billion, and even more difficult to measure, of psychological and esthetic costs.

In the Council on Environmental Quality report (August 1971), the "direct costs of air pollution on both materials and vegetation are estimated at 4.9 billion annually" and on lowering property values 5.2 billion dollars. They conclude:

The annual toll of air pollution on health, vegetation, materials, and property values has been estimated by EPA at more than $16 billion annually—over $80 for each person in the United States. In all probability, the estimates of cost will be even higher when the impact on esthetic and other values are calculated, when the cost of discomfort from illness is considered, and when damage can be more precisely traced to pollutants. Also, the estimates may increase as more is known about the damages of long-term exposure to very low levels of any one pollutant or many in combination. It must be emphasized, however, that these cost estimates only crudely approximate the damages from air pollution.[16]

An example of some of the research presumably used by EPA in their estimates, is the study by Ridker (1967) in St. Louis limited to reducing air pollution effect on housing by lowering sulfation levels somewhat, but in no cases below the "background level." He estimates that 82 million dollars could be saved in St. Louis; invested at 10 percent it would amount to 8 million dollars per year.[17] He goes on to note that "currently available data do not permit a careful comparison between these benefits and the cost of bringing about such a reduction." On the basis of crude estimates, a shift to low sulfur fuels might cut sulfate levels in half and cost 10–15 million per year. This provides more pollution reduction than represented by the 8 million dollars; presumably if the latter were extrapolated the benefit could equal the cost.

Three other points must be borne in mind: (1) property values other than for single family residences would also benefit, (2) other benefits than those from increase in property values would also occur, and (3) cost estimates might also rise because other pollution sources besides sulfur would have to be treated. He concludes: "Considerable work remains before

14. L. D. Zeidberg, R. A. Prindle, and E. Landan, *American Journal of Public Health* 54 (1964) (cited by Lave and Seskin in note 13 above).

15. *Environmental Quality*, Second Annual Report, Council on Environmental Quality, Washington, D.C., August 1971, p. 106.

16. Ibid., p. 107.

17. Ronald G. Ridker and John A. Henning, "The Determinants of Residential Property Values with Special Reference to Air Pollution," *Review of Economics and Statistics* 49 (1967):246–57. For still more discussion see Ronald Ridker, *Economic Costs of Air Pollution* (New York: Praeger, 1967).

an adequate comparison between benefits and costs can be accomplished," although in general he is satisfied that an estimate of air pollution effect on residential property has been devised which can be used with some confidence. The "results however cannot be generalized to other cities and times," although obviously they would be much greater than for just one city.

As just one example since Ridker made his analysis, Commonwealth Edison Company of Chicago is already shifting to sulfur-free coal including dramatically changing sources to new mines for Montana coal brought in by special "unit trains" on a continuous basis a distance of about fifteen hundred miles, five times as far as previously. Other cities are doing the same. The need for sulfur-free fuel, whether coal or oil is dramatically altering major features of the established economic geography of America. Coal mines are being opened in formerly unused areas of America, both for shipping to remote cities and to supply large generating plants in remote areas, as in the Four Corners area of New Mexico, and in Wyoming. These in turn create ecological problems there, and the Indians are complaining.

COSTS OF COUNTERING POLLUTION

Counterposed to the costs of air pollution alone are the costs of cleaning it up, if only partially. Overall estimates for additional expenditures to cut air and water pollution and solid waste disposal run from 32 to 40 billion dollars.[18] Air pollution costs are probably less than half this total. This is probably only an order of magnitude guess; costs might be much higher, depending on the level of treatment desired and the unearthing of new pollutants. On the other hand some of the costs might well be considered non-recurring capital costs not necessary to repeat every year, and some of the costs could well reduce some later production costs—i.e., not be as high net costs as originally.

There is also considerable argument and diversity of opinion over how to arrange for pollution abatement. Cutting all pollution is impossibly costly and unnecessary; likewise the first cut of say 50 or 80 percent in pollution is generally far cheaper than cutting the last or next to last 10 percent, etc., an incremental relationship which economists love to note. Likewise economists dislike uniform standards, pointing out that costs to some industries are greater than to others or other places, and since all pollution need not be cut, uniform standards are costly in relation to benefits.[19] Overlooked may

18. Personal communication, Professor R. J. Christman, College of Engineering, University of Washington.

19. Azrill Teller, "Air Pollution Abatement: Economic Rationality and Reality," in *America's Changing Environment*, ed. Roger Revelle and Hans H. Landsberg (Boston, 1970), pp. 39–55.

be the necessity to have relatively uniform standards in order to even out competition and keep industries from moving unwisely, encouraging adoption of new processes at national and other scales, to say nothing of other equities and enforcement problems. Perhaps a solution is to have reasonable standards, enforceable either by law and/or user changes, and make specific variation from them when industries or the public can make special cases in special places. Certainly local jurisdictions should have the right to higher standards, and to zoning, but some uniform minimum standards as for mobile sources such as construction equipment, trucks, or airplanes are also needed.

The Council on Environmental Quality report (August 1971) has some figures on costs of pollution control to meet certain standards; however, some of these standards do not go into effect until 1975 or 1976 (automotive air pollution, for example), so that the figures are difficult to interpret. Their estimates for total to be invested 1970–75 are 105.2 billions operating and capital expenditures—23.7 air, 38.0 water, 43.5 solid wastes. However the overwhelming bulk of the solid waste costs, they note, are for collection and thus overstate costs (?). On an annualized cost basis, which includes cumulative from 1970–75 and annual capital costs for the last year 1975, they estimate 4.7 billions for air, 5.8 for water, 7.8 for solid wastes or a total of 18.3 billion.[20] However since this only begins to get into the dates for the start of the automotive standards, the figure appears low, although the report does not indicate how low.

In spite of the embryonic nature of the cost of losses and abatements, it seems evident that losses from pollution exceed costs of cure or alleviation. Viewed in still another way a high estimate of 40 billion dollars a year is only about 4 percent of the United States gross national product or 8 percent of total personal income, or something like two hundred dollars per person per year, or more than five hundred dollars per family. On a sustained annual basis it is apt to be less. This has led one economist friend of mine to indicate that the cost is trivial, perhaps too light an assessment.

Philip Handler, president of the National Academy of Sciences, remarked in 1970: "Technology already in being or readily fashioned can rectify the more serious aspects of degradation of air quality and I consider this a relatively temporary, albeit important problem."[21] The problem can be alleviated at no excessive cost, in spite of the fact that much research remains to be done, but to bring it down to levels of pleasant, clean living will be more difficult, especially for the

20. *Environmental Quality* (Council on Environmental Quality), p. 111.

21. Philip Handler, *News Report* (National Academy of Sciences) 20 (August-September 1970):11.

problem of the automobile, where multiple emissions are harder to control than concentrated industrial ones.

Even in this case Henry Ford II indicates (with alarm) that to meet new United States federal emission standards for autos will cost $300 per car and at a new car annual rate of 10,000,000 therefore $300,000,000 a year.[22] (This is about the same cost as incorporating new safety features.) It does not seem too high a cost if the controls are effective.

If the costs of cleaning our air and handling other pollution still appear excessive, then several other possibilities may provide clinching arguments. (1) Health may be much more seriously or dramatically affected than is now provable. (2) Some longer range, wider scale effects on the whole biosphere may legitimately cause alarm. In any case people are often more concerned about the unknown than the known. To quote Professor Handler again: ". . . there is absolutely no likelihood of a significant reduction in atmospheric oxygen; even if all of the known fossil fuel reserves of the world were to be burned tomorrow, this could engender a reduction of only a few percent; our oxygen supply is vouchsafed for the indefinite future. However, the consequences of build-up of atmospheric carbon dioxide are quite uncertain; indeed, one does not know whether, on balance, the results might even be beneficial."[23] (3) As we get more affluent and concerned, our standards will probably rise; what we tolerated in an earlier period, we will not in the future. In spite of increased pollution, we have done just this as witness the rise of air conditioning to escape the debilitating effects of heat (but also creating more outdoor pollution!) and the switch from muddy streams to newly built clear reservoirs or swimming pools for swimming.[24]

OTHER POLLUTIONS BRIEFLY NOTED

Time precludes discussing in detail other pollution already alluded to. *Water* pollution is a serious problem; indeed water treatment and water quality costs are more severe for industry than water supply. Nationally problems of water quality will dominate those of quantity.[25] Serious work on this problem, however, started sooner than for air quality. Indeed, we already have some successes to show, especially Lake Washington in Seattle, a twenty-four-mile long lake which became infested with algae in the 1950s caused by dumping of nutrient rich, primarily residential, treated sewage. This started the lake toward eutropification. Scientific work by Professor W. T. Edmondson of the University of Washington

22. Henry Ford II, *Summary Report Annual Meeting*, Ford Motor Company, 11 May 1971.

23. Handler, *News Report*.

24. Cf. Edward L. Ullman, "Geographical Predition and Theory: The Measurement of Recreation Benefits in the Meramec Basin," *Problems and Trends in American Geography*, ed. Saul B. Cohen (New York: Basic Books, 1967), pp. 124–45.

25. Nathaniel Wollman and G. W. Bonem, *The Outlook for Water*, Resources for the Future (Baltimore: Johns Hopkins University Press, 1971).

was instrumental in establishing the relationship. The sewers were rerouted to Puget Sound, a deeper salt water tidal body, after voters approved expenditures of several hundred million dollars. The citizens in Seattle, as elsewhere, value pure water for recreation. Drinking water supply already came from the mountains and was not at issue. However, the diversion of sewers from the lake plus diversion of drinking water from streams feeding the lake produces problems of insufficient lake water to operate the locks easily during dry summers.

Other lakes, water bodies and streams of the United States are worsening, with industry a principal culprit, as witness Lake Erie and Galveston Bay, to take but two of the worst examples. Study and programs are being pushed on them.

NOISE POLLUTION

An increasing problem of American cities is noise pollution.[26] The problem is not principally from the industrial landscape, but rather from land and air traffic. The airplane is a bad offender as we all know, particularly in the vicinity of airports which are being increasingly surrounded by buildings.

A serious new problem in America is freeway noise. The freeways channel heavy volumes of traffic producing a constant roar, quiet only for a few hours between one and four A.M. This increasing volume of traffic also tends to be on new rights-of-way near residential districts, unlike the old railroads which had already produced an ecology of pollution and adjustments. Commercial or industrial buildings provided a better wall against noise than plantings or trees along freeways. To use a UNESCO phrase: "Green damming" has often only a cosmetic effect on this greatest nuisance of freeways. A further reason for freeway noise is the greater speed of traffic, although the slower traffic, trucks, produce the most individual noise and often are the principal offenders. The noise has been transferred from the industrial-rail landscape to the residential by this medium, hardly progress.

Cities are thus characterized by a background roar primarily of traffic noise with much the same distribution pattern as air pollution. This background is called the ambient noise level; some measure noise against it, feeling it only necessary to keep below this level. However, the constant background noise may turn out to be the real culprit or environmental problem. Industrial processes are noisy and are confined to the industrial site, seriously injuring workers, although laws

26. Some recent studies are: James D. Chalupnik, Ed., *Transportation Noises, A Symposium on Acceptability Criteria* (Seattle: University of Washington Press, 1970); Melville C. Branch, Jr., with R. Dale Beland, *Outdoor Noise and the Metropolitan Environment* (Los Angeles, 1970); Melville C. Branch, Jr., "Outdoor Noise, Transportation and City Planning," *Traffic Quarterly* (April, 1971), pp. 167–88.

curb much of the excess in many countries. Industrial noises, however, are carried to the rest of the city in peculiarly obnoxious form by construction (itself an industrial process), roofing (more for smoke), lawn mowing, and other mechanical activities unnecessarily noisy, and increasingly right into the house by utilities ranging from dishwashers to air conditioners! Many of these noises could be suppressed rather economically. For example, it is estimated that a quiet compressor and jackhammer could be built for only a 10–15 percent increase in cost.[27]

VISUAL POLLUTION

Visual pollution will be only briefly noted. Two aspects of visual pollution related in part to industry, however, deserve singling out: overhead wires, a major problem in some American cities and, even worse, chaotic and uncontrolled signs and billboards, especially along suburban arterial highways. The superhighways however are largely free of signs and wires on their immediate edges since no buildings or access is allowed. Ironically the very signs which are visual pollutants urge us to buy products whose manufacture pollutes the atmosphere and hydrosphere!

THE INTERRELATIONSHIP
BETWEEN POLLUTIONS AND SOLUTIONS

The abatement of one pollution often produces another: water to smoke or vice versa or to solid wastes or back to smoke or water, etc. Thus recycling is urged, but is not always practical. Weight loss is an integral feature of manufacturing, and indeed is the basis for the Weberian or classical theory of industrial location. Weight loss has been cut by more efficient processes and thus technology provides hope for combining efficiency and pollution abatement. One problem is the increased quantity of raw materials resulting from use of lower grade sources as higher grades are used up. Concentration at the source, as in pelletizing iron ore, can shift some of the processing to areas outside cities. Reduced transport costs also make possible shifting to less polluting fuels, as noted before.

Some of the substitutions, however, make one almost despair. For example, burning of clippings, tree prunings, etc., is now prohibited in most American cities. To handle the debris, mobile chipping machines have been developed which produce an enormous amount of noise! The law which

27. *Environmental Quality* (Council on Environmental Quality), p. 102.

bans smoke in this case converts it into noise, hardly environmental progress! If bans against noise were enforced in cities, presumably this could have resulted in the designing of a quieter chipping machine.

The implication from this, of course, is that a balanced program of pollution abatement must be pursued. In America it means particularly that the consideration of environmental implications now required of all federal actions, must especially be applied to environmental actions! This tautology sounds ridiculous, but environmental specialists or bureaucrats often are some of the worst single factor proponents. This fact also may help explain why we are getting so many new forms of pollution! As the Environmental Quality report notes, "As control programs are accelerated in each medium it is likely that extra costs will surface in the other media."[28]

The emphasis in this paper is on pollution of man and primarily of man in cities, where both the concentration of population and pollution is greatest. We should therefore, concentrate our efforts right under our nose. In saving ourselves first we would also be helping the biosphere.

28. *Environmental Quality*, p. 117.

CHAPTER 14   City, Port, and Outport

This paper is an unfinished work. Like most of Ullman's topics, it was one that he had spent years in preparing. In fact, the height of his interest in the topic was in the late 1960s. Nonetheless, he continued to gather information related to the subject right up to the time of his death in 1976, never being completely satisfied that he had sufficient information to proceed into final published form.

"I indeed am still interested in hinterlands of ports, but I have not done much with them in recent years. The whole question of ports is related to the volume of traffic. With vessels becoming larger, small ports are dropping out all over the world, especially with the construction of trucking routes, to say nothing of railroads before. Likewise, there are hinterlands for bulk and densely produced commodities, such as coal or timber, which may be quite small and local, and the port of not much consequence, therefore. The ultimate in this are simply pilings out in the Pacific Ocean, exporting oil from California. The volume is great, but there is no port at all. . . .

"One thing I have been working on recently, is the fact that in many parts of the world, on islands or coastal areas, cities are not on the coast, in spite of Cooley's famous statement in his theory of transportation, that break of bulk is the true city builder effect of transportation. The largest city in the region may indeed be a port, such as

Most of the world's coastal settlements paradoxically are not on the seacoasts. Even if they are known as ports most of them (substantially over 50 percent) are ports only because they are connected with the sea by a special outport, either a separate settlement some miles away, as at Tokyo-Yokohama, or an extension of the city to the coast as at Los Angeles. This separation of city from the sea appears surprising to us in America with our large port metropolises such as Boston, New York, Baltimore, or San Francisco and the strong city building influence that the break-in-bulk between land and water engenders.

This separation is more than the normal separation between port land use and central business districts with their contrasting requirements of freight vs. pedestrian movements, characteristic of all cities to varying degrees. It is a physical separation in many centers greater than that dictated by mere land use requirements. How is this paradox explained? Why such an apparently anomalous development? The answer depends on numerous factors which can be grouped under five headings, two geographical and three historical: (1) situation, the apparent need for the settlement to be centrally located; (2) site, inhospitality or unsuitability of much local terrain along the water's edge; (3) changing technical requirements for deep water; (4) whether the coast was settled from the outside in (as in much of the United States and other areas), or inside out as in much of Europe and Asia and other older settled areas; (5) protection from pirates.

In many cases aspects of all five of the generalized explanations are involved. In other cases still other considerations or special cases exist, since the world naturally has many special arrangements. Nevertheless the statement that most (over half apparently) of coastal settlements are not on the sea coast is true and the explanation is general, not unique.

THE EVIDENCE

First it is necessary to document the statement that coastal settlements are not generally on the coasts, a surprising finding in itself, whatever the reasons therefore.

New York or Havana, but other cities may not be. In some cases, as on the west coast of Mexico, more of the cities are actually a few miles inland, and connected, at least in an earlier period, to the sea by an outport. In this case, the central place function apparently is more important than the port function. . . ." [Letter to Richard G. Boehm, 26 June 1969]

"I have been somewhat surprised that break-of-bulk points are not the principal cities in many regions of the world. With the exception of the east coast, in Latin America neither the largest nor the intermediate size cities are on the coast. For much of the west coast, this may simply be a reflection that the temperate, productive climate is inland. In other parts of the world, cities may be close but not actually on the coast, as in Cuba where the largest cities, such as Havana and Santiago, are coastal, but most of the others are not, even though it is an island. Many parts of the Mediterranean (Rome-Ostia, Pisa-Leghorn), Philippines and particularly Malaya, with Kuala Lumpur, Ipoh, etc., have outport cities. . . ." [Letter to William Alonso, 16 July 1969]

"A main reason I have for writing this, is that I am working on a pet of mine—city and outport around the world. Very briefly I contend that perhaps half of the world's coastal cities are not on the sea coast, but are some miles inland and connected with the sea by an outport. . . . Some reasons for this appear to be: (1) that a central place back of the coast for local trade is more important than a port location only for occasional sea trade, (2) often coasts are marshy, sandy or otherwise unproductive, as well as poor sites for ports in some cases, and (3) subject to pirates, invasions, and malaria. . . ." [Letter to Norman J. G. Pounds, 17 April 1969]

"A topic which I have been playing around with is one which I call "City and Outport," which briefly indicates that perhaps half the world's coastal cities are not on the coast at all but are some distance inland, connected with the coast by an outport, such as Athens-Piraeus, and even more extensive cases in Mexico and Latin America. It's sort of a triumph of central place over break-of-bulk plus unattractive physical conditions along coasts in many cases, as well as pirates. I hope to write you more later. . . ." [Letter to William Douglas Warren, 17 July 1972]

APPENDIX: An Autobiographical Statement

To
Robert B. McNee,
Department of Geography,
University of Cincinnati
August 25, 1971

Dear Bob:

I am flattered that you want biographical and other material about me, but I do not respond primarily for this reason, but rather because I too am much interested in biography. It is a favorite form of reading for me.

I enclose a curriculum vita together with many reprints and some unpublished material which will offer you some clue about myself, although I do not know how much all the published material really represents me. They do, but they also represent influences and *demands*. I believe that everything I have done I have attempted to do as a geographer and furthermore have always attempted to have an *idea* in mind.

Unlike many geographers I have always been a geographer. I became one even before entering high school. When I was 8 or 9 years old my mother gave me a ten cent store diary. In it was the populations of the cities of the country. I could not read four or more digit figures and so I asked my mother what they meant. She told me and I proceeded to memorize the approximate populations of the cities and towns of the United States. A year later I asked, as a Christmas present, to get a wall map of the United States. This early interest in statistics and maps is remarkably parallel to what Walter Christaller describes of his youth. When I said I was a born geographer I really meant an economic geographer, since I was concerned with industries, cities, and transportation, probably because I grew up in cities and towns, in contrast to Carl Sauer, who was interested in agriculture and the natural environment.

Retrospectively a partial reason I became a geographer, I suppose, is because my family moved around. At the age of 7 we moved from Pittsburgh out to Iowa City and at the age of 13 I went to Rome, Italy for a year and then to Chicago. Perhaps equally interesting to me was transportation but it was transportation as a geographer, I suppose. I never intended to become an academic geographer but rather wished to work in transportation, but since I had no relatives in transportation I never got a job. However in college I did serve as a travel agent and knew everything about that business, including the size of all the ships, the schedules of all the trains, etc. It is amazing how much information one can acquire if one is interested and young. As a result in my geography courses in college I often tended to answer questions backward—because I knew, not because I reasoned the answer out. I had a good memory for what interested me. However, this has not been my major method of operating. I realize that I have always had concepts in mind and tend to develop answers based on chasing the concepts. You don't find something unless you are looking for it.

In elementary school I did not do particularly well in my studies but always managed to get 100 in geography; however, I have always felt that I was self taught, which does give one a certain originality, but also has its weaknesses.

In my geographical training, in addition to the descriptive approach, I was also exposed to a lot of work in geology, and the natural environment. While I was not particularly interested in that subject matter, the approach was similar to that which a geographer would apply to economic features; likewise I was never interested in Roman ruins, museums, etc., which I was dragged through innumerable times in Europe. In this respect my interest was distinct from my father's, just as it was distinct from my teachers. In fact, after my first two quarters at the University of Chicago, I almost decided not to become a geographer, but then I took a course with Professor Charles C. Colby and his interest and mine coincided much more.

As a graduate student at Chicago I also contacted on my own Louis Wirth, professor of Sociology, and had the run of his office. (Interestingly, the Chicago geographers and sociologists had never met, even though both espoused human ecology.) It was Wirth who asked me to prepare the article on "A Theory of Location for Cities." I even gave some lectures to his course in Human Ecology. However, I have never had a course in sociology! I also have had several long conversations with Rupert Vance at North Carolina and also knew Howard Odum.

As indicated, I have been exposed to many influences: (1) one of them, I am sure was my father who was a distinguished classical scholar and taught at the University of Chicago and other places. Major league performance seemed to me to be the only goal, however much one's weaknesses prevented achieving it! (2) I went to Chicago and Harvard which reinforced this approach. (3) I taught in an Economics Department at Washington State from the age of 23–25 and imbibed some theory from colleagues although I never liked conventional economics. I also had extensive discussions with Walter Isard at Harvard from 1946–50. (4) I was also in a position, probably, to know more geographers in the period 1940–1955 than anyone else in my age group, and had extensive discussions with Hartshorne, James, Stephen Jones, and of course Chauncy Harris and Ed Ackerman, who were my contemporaries in graduate school. I was also very close to Whittlesey, since for quite a time we were the two geographers at Harvard. I remember discussing with him Hartshorne's *Nature of Geography*, when it came in manuscript to Whittlesey in 1938. (5) As noted above, I came under the influence of the descriptive or deadpan approach to American geography in my college years before World War II and this left some mark, although I always tended to rebel against that, and therefore espoused theory and quantitative approaches at an early age. However theory in terms of simple abstract statement did not particularly appeal to me nor do I like mathematics nor wish to be a quantitative technician. I like to use these for my own substantive, geographical ideas. (6) I have probably spent more time than practically any other geographer in trying subtly to prove the value of geography (and myself too, I suppose) by doing jobs outside the field. I early came to the conviction that the main need in geography was to prove that it could be useful and meet the standards of other disciplines. This is a rather interesting point of view for someone like me who is intensely interested in geography qua geography, or geography as I define it, primarily spatial relations. However, I felt it was a duty to do this and I suppose I have always been influenced by this sort of sense of responsibility even including public service and trying to reform the world, although I do not know whether this is my main motivation. I prefer to do things as opposed to talking about them. (7) Another influence on me was my work in OSS where I had to prove myself to other disciplines and where geography unfortunately did not have too good a reputation. (Sherman Kent called them unemployables, indicating his prejudice.)

I remember Isaiah Bowman once telling me that the age of 30–35 was his most productive period. For my generation such was not the case; we were at war then and thus our most productive period was directed toward applied ends. I did, however, manage to develop some new techniques; in a war and in intelligence everybody started from scratch, so those of us who were young and had just finished our Ph.D.'s were in a position to rise rapidly, which we did. The same was true in England. Much of the work of course was not geographical and during the later part of the war I served as a member and later director of a joint intelligence study board which had a great influence in raising the level of intelligence work, and I suppose represented my main contribution as well as the main contribution of geography up to that time, although the reports unfortunately were classified as confidential even though I almost had them declassified at the end of the war. At any rate I left the war very tired, even though I had planned to stay on in the government and indeed had a permanent job with the Maritime Commission. However, I went to Harvard instead on a joint appointment in Planning and Geography.

(8) After three years geography was essentially abolished because of local, internal Harvard reasons. I remember going to Edward S. Mason, Professor of Economics and Dean of the Graduate School of Public Administration at Harvard who had been in OSS and he said: "Ed, I will do anything for you but nothing for geography." This was the situation in the field. When geography was abolished considerable fuss was raised (not entirely spontaneously) and eventually the Provost appointed a committee to investigate geography. For two years we met with me as the only geographer on a committee, highly skeptical, of peers from other disciplines. Leading geographers testified and most of them did not make much impression, but we persevered and at the end of two years the committee unanimously recommended the creation of a separate department of geography, something which Harvard had never had. In other words we achieved a moral victory but no money was ever forthcoming to implement it. However, this defense also made an impression on me and again influenced me to try to prove geography's worth. Initially I was asked to stay on in the Planning Department and promoted to Associate Professor; however, I preferred to be a geographer and when the opening at Washington occurred I came out here.

I have continued to pursue two sorts of lines of geographic inquiry; one, more or less to my own liking, as represented by "Rivers as Regional Bonds," and the other, attempts to prove the worth of geography which resulted in two distinctive inventions, the minimum requirements approach to the Economic Base, and even more applied, the technique for estimating recreation benefits during my two year study of the Meramec Basin. Neither of these inventions particularly interested me but they reflect a geographical approach.

Earlier, in part influenced by the need for theory, I had devised a central place theory and then discovered that Walter Christaller had already developed it and had carried it much further than I ever had in my mind. I believe I published something on how I came to that theory (including my conversation with August Lösch at Harvard in 1938), in the Lund Symposium on Urban Geography, c. 1960, after the Stockholm IGU Meeting.

The term spatial interaction I coined when I went to a meeting of the Western Political Science Association at Pullman, Washington in 1952, where the speaker before me, talking about Sociology, defined it as the study of social interaction. I thought, "Aha, by the same token geography can be defined as the study of spatial interaction." Thus I have been subject to interdisciplinary inspiration also. However, another point that has always intrigued me is that ideas seem to float independently in the air in people's minds who have no contact with each other; such was the case with Central Place Theory for Christaller, Lösch, and myself in a modest way. It is often easier to have an original idea in English than to read it second hand in German, I say, or from Economics or some other abstruse subject!

I have also done a lot of administrative work, in part reflecting a desire to participate. Now, however, I am consciously trying to do my own thing as represented by my Space/Time manuscript, and in a totally different context a notion which I call "City and Outport" which briefly holds that perhaps half the world's coastal cities are not on the coast at all, but rather are a few miles inland and may be connected with the coast by an outport.

However, I get detoured into other things such as reading a paper on "The Problems of the Industrial Landscape" to a UNESCO meeting in June in Bonn dealing with "Man and the Biosphere" in which I put forth my notions, particularly on air pollution over cities, a subject of interest to me. Another detour is represented by the reprint on Higher Education which I felt someone should speak out on to indicate the dismal fate in store for the University of Washington if the legislature socked us. They did. I'm still fighting—just asked to be on a statewide committee. For years I have planned to get out a book consisting of reprints of some of my articles with some changes plus some new items and a general introduction.

The above rambling reply merely scratches the surface and I am sure that I have left out much, but perhaps it will give you a start. Let me know if you have any advice for me. Remember I need to write the introduction to the book of my collected works!

Let me congratulate you again for having such a stimulating idea. I give a course, more or less, on what geography is for entering graduate students and undoubtedly could use some of your insights. Best regards.

Bibliography

SELECTED PUBLICATIONS OF EDWARD L. ULLMAN

Monographs

1943   *Mobile: Industrial Seaport and Trade Center* (Chicago: Department of Geography, University of Chicago). 167 pp. Lithoprinted.

1957   *An Economic Analysis of Philippine Domestic Transportation* (with R. O. Shreve, H. E. Robison, R. K. Arnold, J. W. Landregan, and J. A. McCunniff). Prepared for the National Economic Council of the Philippines. (Menlo Park, California: Stanford Research Institute). 7 vols. 1129 pp.

1957   *American Commodity Flow: A Geographical Interpretation of Rail and Water Traffic Based on Principles of Spatial Interchange* (Seattle: University of Washington Press). 215 pp. Reprinted 1959, 1967.

1961–62 *The Meramec Basin: Water and Economic Development* (with Ronald R. Boyce and Donald J. Volk). Report of the Meramec Basin Research Project (St. Louis, Missouri: Meramec Basin Research Project, Washington University). 3 vols.

1969   *The Economic Base of American Cities: Profiles for the 101 Metropolitan Areas over 250,000 Population based on Minimum Requirements for 1960* (with Michael F. Dacey and Harold Brodsky) (Seattle: University of Washington Press). 112 pp. Revised and enlarged, 1971. 118 pp. (University of Washington, Center for Urban and Regional Research, monograph no. 1.)

Articles, Chapters, and Reports

1936   "The Historical Geography of the Eastern Boundary of Rhode Island," *Research Studies of the State College of Washington*, Pullman, vol. 4, no. 2, pp. 67–87.

1939   "The Eastern Rhode Island-Massachusetts Boundary Zone," *Geographical Review* 29:291–302.

1941   "A Theory of Location for Cities," *American Journal of Sociology* 46:853–64.

1945   "The Nature of Cities" (with Chauncy D. Harris), *Annals of the American Academy of Political and Social Science* 242:7–17.

1949   "The Railroad Pattern of the United States," *Geographical Review* 39:242–56.

1951   "Rivers as Regional Bonds: The Columbia-Snake Example,"
        Geographical Review 41:210–25.

1953   "Human Geography and Area Research," *Annals, Association
        of American Geographers* 43:54–66.

        "Transportation Geography" (with Harold M. Mayer), chap-
        ter 13 in *American Geography, Inventory and Prospect*, edited by
        Preston E. James and Clarence F. Jones (Syracuse, New York:
        Syracuse University Press for the Association of American
        Geographers), pp. 310–32.

        "Amenities as a Factor in Regional Growth," *Geographical Re-
        view* 44:119–32.

        "Geography as Spatial Interaction," in *Interregional Linkages.
        Proceedings of the Western Committee on Regional Economic
        Analysis*, Social Science Research Council, Berkeley, Califor-
        nia, pp. 63–71.

1955   "Die wirtschaftliche Verflechtung verschiedener Regionen
        der USA betrachtet am Güteraustausch Connecticuts, Iowas
        und Washingtons mit den anderen Staaten," *Die Erde,
        Zeitschrift der Gesellschaft für Erdkunde zu Berlin*, Band 7, Heft 2,
        pp. 129–64.

1956   "The Role of Transportation and the Bases for Interaction," in
        Man's Role in Changing the Face of the Earth, ed. William L.
        Thomas, Jr. (Chicago: University of Chicago Press), pp. 862–
        80.

1958   "Regional Development and the Geography of Concentra-
        tion," *Papers and Proceedings of the Regional Science Association*
        4:179–98.

1959   "Lo sviluppo regionale e la geografia della concentrazione con
        particolare riguardo agli Stati Uniti, *Bollettino della società geo-
        grafica Italiana*, serie 8, vol. 12, pp. 319–44.

        "Sources of Support for the San Francisco Bay Area Economic
        Base," in *Future Development of the San Francisco Bay Area
        1960–2020*, U.S. Department of Commerce, Office of Area
        Development (Washington, D.C.: U.S. Government Printing
        Office), pp. 34–40.

1960   "Trade Centers and Tributary Areas of the Philippines," *Geo-
        graphical Review* 50:203–18.

        "Geographic Theory and Underdeveloped Areas," in *Essays
        on Geography and Economic Development*, ed. Norton Ginsburg
        (Chicago: University of Chicago, Department of Geography,
        Research Paper No. 62), pp. 26–32.

        "The Minimum Requirements Approach to the Urban
        Economic Base" (with Michael F. Dacey), *Papers and Proceed-
        ings, Regional Science Association*, vol. 6, pp. 175–94. Also pub-
        lished in 1962. *Proceedings of the IGU Symposium in Urban Geog-*

raphy, Lund 1960, ed. Knut Norborg, *Lund Studies in Geography*, series B, no. 24 (Lund: C. W. K. Gleerup), pp. 121–43.

1962   "An Operational Model for Predicting Reservoir Attendance and Benefits: Implications of a Location Approach to Water Recreation" (with Donald J. Volk), *Papers of the Michigan Academy of Sciences, Arts, and Letters* 47:473–84.

"The Nature of Cities Reconsidered," *Papers and Proceedings, Regional Science Association* 9:7–23 (Presidential Address to the Regional Science Association).

1967   "Geographical Prediction and Theory: The Measurement of Recreation Benefits in the Meramec Basin," in *Problems and Trends in American Geography*, ed. Saul B. Cohen (New York: Basic Books), pp. 124–45.

1968   "Minimum Requirements after a Decade: A Critique and Appraisal," *Economic Geography* 44:364–69.

"The Primate City and Urbanization in Southeast Asia: A Preliminary Speculation," SEADAG (Southeast Asia Development Advisory Group) *Papers on Problems of Development in Southeast Asia*, no. 31 (New York: The Asia Society). 9 pp.

1974   "The City and Environmental Quality, Especially Air Pollution Sources and Costs," *Der Mensch und die Biosphäre, Proceedings of UNESCO Conference, No. 20, on Man and the Biosphere* [1971] (Pullach bei München: Verlag Dokumentation), pp. 10–27.

"Space and/or Time: Opportunity for Substitution and Prediction," *Transactions of the Institute of British Geographers*, no. 63, pp. 125–39.

WRITINGS ABOUT EDWARD ULLMAN

Caralp, Raymonde
1976   "Edward L. Ullman, 1913–1976," *Annales de géographie*, no. 472 (November–December), p. 725.

Della Valle, Carlo
1976   "Edward L. Ullman," *Bollettino della società geografica italiana*, vol. 5, nos. 10–12 (October–December), pp. 534–35.

Eyre, J. D., ed.
1978   *A Man for All Regions: The Contributions of Edward L. Ullman to Geography*. Papers of the Fourth Carolina Geographical Symposium (Chapel Hill, N.C.: University of North Carolina, Studies in Geography No. 11). 158 pp.

Harris, Chauncy D.
1977   "Edward Louis Ullman, 1912–1976," *Annals of the Association of American Geographers*, vol. 67, no. 4 (December), pp. 595–600.

Index